Sorin Dan Anghel

Thermally Non-aggressive Atmospheric Pressure Plasma

AF138645

Sorin Dan Anghel

Thermally Non-aggressive Atmospheric Pressure Plasma

Generation, characterization, applications

LAP LAMBERT Academic Publishing

Impressum / Imprint

Bibliografische Information der Deutschen Nationalbibliothek: Die Deutsche Nationalbibliothek verzeichnet diese Publikation in der Deutschen Nationalbibliografie; detaillierte bibliografische Daten sind im Internet über http://dnb.d-nb.de abrufbar.

Alle in diesem Buch genannten Marken und Produktnamen unterliegen warenzeichen-, marken- oder patentrechtlichem Schutz bzw. sind Warenzeichen oder eingetragene Warenzeichen der jeweiligen Inhaber. Die Wiedergabe von Marken, Produktnamen, Gebrauchsnamen, Handelsnamen, Warenbezeichnungen u.s.w. in diesem Werk berechtigt auch ohne besondere Kennzeichnung nicht zu der Annahme, dass solche Namen im Sinne der Warenzeichen- und Markenschutzgesetzgebung als frei zu betrachten wären und daher von jedermann benutzt werden dürften.

Bibliographic information published by the Deutsche Nationalbibliothek: The Deutsche Nationalbibliothek lists this publication in the Deutsche Nationalbibliografie; detailed bibliographic data are available in the Internet at http://dnb.d-nb.de.

Any brand names and product names mentioned in this book are subject to trademark, brand or patent protection and are trademarks or registered trademarks of their respective holders. The use of brand names, product names, common names, trade names, product descriptions etc. even without a particular marking in this works is in no way to be construed to mean that such names may be regarded as unrestricted in respect of trademark and brand protection legislation and could thus be used by anyone.

Coverbild / Cover image: www.ingimage.com

Verlag / Publisher:
LAP LAMBERT Academic Publishing
ist ein Imprint der / is a trademark of
OmniScriptum GmbH & Co. KG
Heinrich-Böcking-Str. 6-8, 66121 Saarbrücken, Deutschland / Germany
Email: info@lap-publishing.com

Herstellung: siehe letzte Seite /
Printed at: see last page
ISBN: 978-3-659-50559-1

Copyright © 2013 OmniScriptum GmbH & Co. KG
Alle Rechte vorbehalten. / All rights reserved. Saarbrücken 2013

Table of Contents

Preface

It is well known the interest of scientists to develop new thermally non-aggressive atmospheric pressure plasma sources. The interest in this topic is dictated by a potential economic benefit from numerous non-thermal plasma technologies: plasma assisted chemical vapor deposition, etching, polymerization, protective coating deposition, toxic and harmful gas decomposition, sterilization and decontamination, electromagnetic wave shielding, polymer surface modifications, nanoparticle synthesis, and so on. These new technologies can be now applied on materials that cannot be treated at low-pressure or cannot support temperatures much higher than the ambient temperature. Such plasmas are placed in the category of non-thermal plasmas, defined as having the gas temperature lower than the combustion temperature $(T_{gas} < T_{combustion} = 2300$ K). When atmospheric pressure plasmas are used for surfaces treatment or for material processing, three main requirements must generally be fulfilled: to be homogeneous over large treatment areas, to have low temperature (close to the ambient temperature) so the plasma to be thermally non-aggressive with the treated materials and to be capable to generate chemically active species.

Atmospheric pressure plasmas overcome the disadvantages of vacuum operation but their generation presents difficulties because of the very high electric voltages (kV) needed for gas breakdown. Moreover, the difficulties imposed by the necessity of adapting the output impedance of the waveform generator to the plasma impedance must be passed. Commercial plasma generators solve these problems by interconnecting a matching network between the waveform generator and the plasma torch. My idea was to solve both problems by generating the plasma as intrinsic part of an *LRC* series resonant circuit. The particular characteristics of the series resonant circuit represent the development basis of some atmospheric pressure plasma generators. Plasmas with powers ranging from hundreds of mW to several tens of watts were generated at frequencies ranging from tens of kHz to a few MHz.

The purpose of this book is to provide students and researchers with a comprehensive guide for construction their own atmospheric pressure plasma generators at very low cost. All high voltage generators and plasma chambers described here were constructed with few resources. The generated plasmas were characterized and tested for bacterial deactivation, surface energy modification, organic dye degradation and nanoparticle synthesis.

This book is a short review of my contributions to the study of thermally non-aggressive atmospheric pressure plasma. Most of the obtained results were published in works whose single author, first author or corresponding author I was (see references 1-9, 30 and 50).

I would like to thank the Elsevier Publisher for permission to use text blocks and figures from ref. [6], [7] and [50] and the IOP Publisher for permission to use figures from ref. [2], [3] and [30]. I would also thank my sons, Dan and Mihai for attentively revision of the text and for the help in the editing work.

Finally, I gratefully acknowledge the loving support of my wife, Ági.

Sorin D. Anghel

Cluj-Napoca, 2013

Abbreviations

AC	alternating current
AES	atomic emission spectroscopy
AP	atmospheric pressure
APP	atmospheric pressure plasma
APPJ	atmospheric pressure plasma jet
CCP	capacitively coupled plasma
DC	continuous current
DBD	dielectric barrier discharge
LTE	local thermodynamic equilibrium
MOSFET	metal-oxide-semiconductor field-effect transistor
pk-pk	peak to peak
RF	radiofrequency
rms	root mean square
TE	thermodynamic equilibrium
TEM	transmission electron microscopy

1 Basics [1,4,9,30]

1.1
Cold non-thermal plasma

Plasma is a quasi-neutral and very complex system consisting of a large number of species: electrons, positive and negative ions, free radicals, gas atoms and molecules in the ground or excited states, photons and electromagnetic radiations. The most common way to generate laboratory plasma is by placing the plasma chamber under the action of an electric field from which the gaseous medium absorbs the energy necessary to pass in plasma state. Taking into consideration the energy transferred to the plasma (and implicitly the energy of the constituent particles) plasma can be considered as the fourth state of the matter, apart from the solid, liquid, and gas states.

The temperature is a physical parameter associated with the energy of a system. Plasma can be considered a complex system consisting of two categories of species: the light species (electrons) and heavy species (neutral atoms or molecules, ions and neutral molecular fragments). Because of their higher mobility, the electrons gain more energy than the ions from the electric field, generating and maintaining the plasma state by transferring a part of their energy to other species via elastic or inelastic collisions. The absorbed energy is found in plasma in various forms: mechanic (translational), excitation, ionization, dissociation, radiation. Each of these energies can be associated with a temperature: the *translational ion* and *electron temperatures,* T_{ion} and T_e, which characterize the translational energies of the ions and electrons; the *excitation temperature*, T_{exc}, which characterizes the energy of the excited particles in the plasma; the *ionization temperature*, T_{ion} and the *dissociation temperature*, T_d, which characterize the energy of ionization and dissociation, respectively; and the *radiation temperature*, T_r, which characterizes the radiation energy. Generally it is accepted as the *gas temperature*, T_{gas}, the temperature which characterizes the translational energy of heavy particles (atoms, ions, molecules). The state in which all these temperatures are equal is a state of TE. Plasmas in which TE is achieved in volumes of order of the mean free path length are called LTE plasmas. Most of laboratory plasmas are not in thermodynamic equilibrium. In these plasmas all temperatures are different.

Generally, plasmas can be divided into two main groups: the *high temperature* or *fusion plasmas* and the *low temperatures plasmas* or *gas discharges*. The high temperature plasmas are in TE state. The low temperature plasmas can be in LTE state or non-equilibrium state

The term of *non-thermal plasma* denominates plasmas which are not in thermodynamic equilibrium. In these plasmas either the electron and ion temperatures are different (usually $T_e \gg T_i$), or the velocity distribution of one or more species does not follow a Maxwell-Boltzmann distribution. A particular type of non-thermal plasma is called *cold plasma* whose gas temperature at atmospheric pressure is near room temperature.

Many technological applications of plasmas (etching, thin film deposition, surfaces treatment) are based on gas discharges generated at pressures lower than the atmospheric pressure. Because APPs overcome the disadvantages of vacuum operation, these plasmas, particularly the non-thermal plasmas, have been given increased attention over the last decades. They offer high excitation selectivity and energy efficiency in plasma chemical reactions. They are also sources of UV, VIS and IR radiations, free radicals (such as O and OH) and active species that can play important roles in various techniques. Many kinds of low power and very low power APP sources for biomedical, environmental and technological applications were designed. The most important of them can be considered: plasma needle, plasma pencil, miniature pulsed glow-discharge torch, open-air hollow slot micro-plasma, one atmosphere uniform glow-discharge plasma, resistive barrier discharge, dielectric barrier discharge, and atmospheric pressure plasma. Among their numerous applications, the following must be mentioned: spectral sources for atomic emission and fluorescence spectroscopy, gas treatments and material processing (deposition of thin films, etching, ion implantation, surface activation of polymers, plasma polymerization, ashing, oxidation and surface hardening), ozone generation, environmental and biomedical applications and particle sources.

The knowledge of the plasma parameters allows us to have information on the elementary processes in the plasma. It is also well known that the atmospheric pressure laboratory plasmas are under non-thermal equilibrium state. The evaluation of plasma temperatures and electron number density is important in understanding the interaction processes. The study of the emission spectrum is a very useful tool for thermal characterization of plasmas. Moreover, this tool represents a non-intrusive and non-perturbing method. By ascribing emission lines to atoms and molecules, the atomic and molecular species and their generation and excitation mechanism in different plasma zones were identified and studied. The relative intensities of the atomic and molecular emission lines were used to estimate the plasma temperatures based on Boltzmann or Saha plot or by finding the best fit of the measured spectra with the synthetic spectra. The hydrogen emission line H_α was used to calculate the electron number density based on the Stark broadening.

The difficulty of sustaining an APP leads to high voltages ($>10^3$ V) necessary to gas breakdown, which can produce the transition of the

7

luminescent silent discharge to an electric arc. In our researches we have solved these two problems (avoidance of arcing on the one hand and generating high voltages on the other hand) by using a single electrode to generate and sustain the discharge or by placing dielectric plates on the electrodes. The powered electrode is connected at the end of a coil which is a component part of an *LRC* series resonant circuit. The presence of the plasma as a part of the resonant circuit assures a good transfer efficiency of the power from the generator to the discharge.

1.2
Series-resonant circuit

One necessary condition for generating an electrical discharge is for the voltage between the electrodes of the plasma chamber to be at least equal to the gas breakdown voltage. When the plasma is generated at atmospheric pressure the breakdown voltage could be of several kilovolts, depending on the working frequency of the applied electric field and on the kind of plasma gas. The main particularity of all our non-thermal plasma generators is that for generating such high voltages we used the particular characteristics of *LRC* series circuits. Such a circuit is shown in figure 1.1.

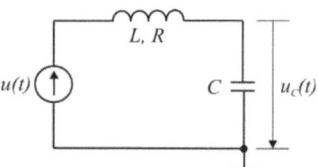

Figure 1.1 – *LRC* series circuit. [30]

Where L and R are the inductance and the loss resistance of the coil, C is the capacitance of the capacitor and $u(t) = U\sin\omega t$ ($\omega = 2\pi f t$, f - frequency of the AC power supply). The impedance of circuit:

$$Z = \sqrt{R^2 + \left(\omega L - \frac{1}{\omega C}\right)^2} \qquad (1.1)$$

depends on the working frequency f, and it has a maximum when $\omega L = 1/\omega C$. This particular frequency depends only on the characteristics of the coil and capacitor and is given by $f_o = 1/2\pi\sqrt{LC}$. At this particular frequency, called the resonance frequency, the circuit has a purely resistive behavior ($Z = R$), the current flowing through circuit is maximum and the voltages across the inductive and capacitive reactances are also maximum but their phasorial sum is zero. Generally, the relationship between the voltage across the capacitor and the DC supply voltage is expressed by a factor k:

$$k = U_C/U = 1/\sqrt{\left(1 - \left(\frac{f}{f_o}\right)^2\right)^2 + \frac{R^2}{\omega_o^2 L^2}\left(\frac{f}{f_o}\right)^2} \qquad (1.2)$$

It depends on the working frequency and, as it can be seen in figure 1.2, it has a maximum at f_o (the resonance frequency for the exemplified circuit is 40 kHz). This value of k (k_{max}) is called the quality factor, Q of the LRC circuit.

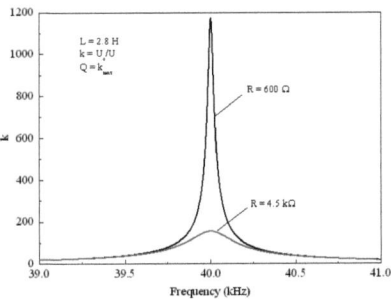

Figure 1.2 - k factor dependence on the oscillating frequency for the same circuit with two different loss resistances. [30]

The quality factor can be calculated based on the characteristics of the circuit elements with the expression $Q = \sqrt{LC}/R$ and it is strongly dependent on the loss resistance of the coil (particularly) and of the whole circuit (in general). The higher the quality factor, the higher the voltage across the reactive parts will be at the resonance frequency. This specific characteristic of resonant circuits was "speculated" for generating atmospheric pressure plasmas. More specifically, the series resonant circuit can be the secondary winding of a transformer and one of the electrodes of the plasma chamber is connected to one end of the secondary coil. When the plasma is generated in an LRC series circuit and when the plasma can be modeled as a simple resistor, the quality factor of the circuit decreases drastically. Figure 2.1 shows one resonance curve for an LRC circuit in the absence of the plasma ($R = 600\ \Omega$) and another one in the presence of the plasma ($R = 4.5\ k\Omega$).

The fact that under the resonant condition the voltages across both of the reactive circuit elements of the series LRC circuit are Q times higher than the supply voltage that was used by us to obtain the sufficiently high voltages for generating two kinds of APPs: inductively coupled plasma (ICP) and capacitively coupled plasma (CCP). This classification is based on the physical mechanisms by which the energy is transferred from the electromagnetic field to the plasma. ICP absorbs almost all its energy from the electromagnetic field via the electromagnetic induction phenomena, like a voltage transformer, while CCP absorbs its energy from the electric component of the electromagnetic field, like a real capacitor. In both cases the plasma acts as an intrinsic part of a resonant circuit. The present work presents only CCP generators for thermally non-aggressive plasmas.

1.3
Ionized gas in a sinusoidal electric field

1.3.1
Plasma conductivity

The process of the electric breakdown of a gas under the action of a bipolar variable voltage depends on the electric field frequency and on the gas pressure. Those who are mainly "responsible" for the process are the electrons because their mobility is much higher than the ions mobility. They absorb the energy from the electric field and then transfer it to neutral atoms and molecules, generating new charged particles. The process takes place in a different way than in continuous current because the electric field polarity changes periodically. At high pressures or atmospheric pressure the collision frequency of the electrons with neutrals is very important. Hereinafter it will be shown that the transfer efficiency of the energy from the electric field to the ionized gas is maximum when the collision frequency and the plasma frequency are equal.

Consider an electron moving in an sinusoidal electric field, $\vec{E} = \vec{E}_o e^{j\omega t}$. The motion takes place in the presence of electron collision with neutrals (atoms and/or molecules). The equation of motion of the electron is given by:

$$m_e \frac{d\vec{v}_e}{dt} = -e\vec{E}_o e^{j\omega t} - v_c m_e \vec{v}_e \tag{1.4}$$

where:

$\vec{v}_e$ - electron drift velocity

e, m_e - electron charge an mass

$\vec{E}_o$ - electric field amplitude

$\omega = 2\pi f$ - angular frequency of the electric field

v_c - momentum transfer frequency between electrons and gas atoms and/or molecules.

Equation (1.4) can be rewritten as:

$$\frac{d\vec{v}_e}{dt} + v_c m_e \vec{v}_e = -\frac{e\vec{E}_o}{m_e} e^{j\omega t} \tag{1.5}$$

Its solution is a linear combination of the homogeneous equation solution and a particular solution of the non-homogeneous equation. The solution of the homogeneous equation:

$$\frac{d\vec{v}_e}{dt} + v_c m_e \vec{v}_e = 0 \qquad (1.6)$$

is: $\vec{v}_{eo} = \vec{k}_o e^{-v_c t}$ ($\vec{k}_o$ - constant) $\qquad (1.7)$

A particular solution of equation (1.5) has to reflect the sinusoidal motion of the electron, so it is chosen as:

$$\vec{v}_{e1} = \vec{k}_1 e^{j\omega t} \qquad (\vec{k}_1 \text{ - constant}) \qquad (1.8)$$

From (1.5) and (1.8) the expression:

$$\vec{k}_1 = -\frac{e\vec{E}_o}{m_e} \frac{1}{v_c + j\omega} \qquad (1.9)$$

is obtained for $\vec{k}_1$, and the particular solution will be:

$$\vec{v}_{e1} = -\frac{e\vec{E}_o}{m_e} \frac{1}{v_c + j\omega} e^{j\omega t} \qquad (1.10)$$

The general solution of the equation (1.5) can be written as follows:

$$\vec{v}_e = -\frac{e\vec{E}}{m_e} \frac{1}{v_c + j\omega} + \vec{k}_1 e^{-v_c t} \qquad (1.11)$$

Knowing that at high pressure, particularly at atmospheric pressure, the collision frequency is very high, the second term of the equation (1.11) drastically decreases in time and the expression of the velocity of an electron moving in a sinusoidal electric field can be well approximated by:

$$\vec{v}_e(t) \cong \frac{-e\vec{E}}{m_e(v_c + j\omega)} \qquad (1.12)$$

Having in view the well-known expressions:

12

$$\vec{v}_e = -\mu_e \vec{E} \tag{1.13}$$

$$\vec{j}_e = -n_e e \vec{v}_e = n_e e \mu_e \vec{E} \tag{1.14}$$

$$\vec{j}_e = \sigma_e \vec{E} \tag{1.15}$$

where n_e is the electron number density and μ_e is the electron mobility, the electrical conductivity, σ_e, can be expressed as:

$$\sigma_e = \frac{n_e e^2}{m_e} \frac{v_c}{v_c^2 + \omega^2} - j \frac{n_e e^2}{m_e} \frac{\omega}{v_c^2 + \omega^2} = \sigma_{er} + j\sigma_{ei} \tag{1.16}$$

Based on this expression, the ionized gas can be modeled as an impedance consisting of a dissipative (resistive) component and a reactive component whose ratio is v_c/ω. The resistive component is "responsible" for the energy transfer from the electric field to the ionized gas. The transferred power density can be calculated as it follows:

$$p_\approx = \frac{1}{T} \int_0^T j_{er} E_r dt = \frac{1}{T} \int_0^T \sigma_{er} E_r^2 dt = \frac{n_e e^2 E_o^2}{2 m_e} \frac{v_c^2}{v_c^2 + \omega^2} \tag{1.17}$$

where j_{er} means the electron current density and $E_r = E_o \cos\omega t$ represents the real component of the electric field. The expression (1.14) suggests us two conclusions: (**a**) in the absence of collisions ($v_c = 0$) of the electrons with neutral particles (atoms and/or molecules) no energy is transferred from the electric field to the ionized medium; and (**b**) the transferred energy is maximum when the collision frequency is equal to the oscillating frequency of the electric field.

By comparing the transferred power density from the sinusoidal electric field (14) with that transferred from a DC electric field, $E_=$:

$$p_= = j_e E_= = \frac{n_e e^2}{m_e v_c} E_=^2 \tag{1.18}$$

the concept of *effective field* can be introduced, expressed as:

$$E_{ef}^2 = \frac{1}{2} \frac{v_c^2}{v_c^2 + \omega^2} E_o^2$$

(1.19)

1.3.2

Plasma permittivity

We consider a neutral ($n_i = n_e = n_o$) plasma layer having a finite width, d. Because of much higher mobility of electrons than that of ions, a displacement of the electrons with respect to the ions by a small distance $x << d$ can take place. According to Gauss' law, the strength of the electric field generated by the two surface charges (electrons and ions clouds) will be:

$$E(x) = \frac{e n_o}{\varepsilon_o} x$$

(1.20)

The force acting on the electrons,

$$F_e(x) = -\frac{e^2 n_o}{\varepsilon_o} x$$

(1.21)

is an elastic kind one. The angular oscillating frequency of the electron cloud,

$$\omega_{pe} = \frac{e^2 n_o}{\varepsilon_o m_e}$$

(1.22)

represents the fundamental characteristic of a plasma, called *the electron plasma frequency*. Naturally, the ion cloud will oscillate with a frequency

$$\omega_{pi} = \frac{e^2 n_o}{\varepsilon_o m_i}$$

(1.23)

which is much lower than the oscillation frequency of the electron cloud, so it can be approximated that the electron plasma frequency represents even the *natural frequency of the plasma*, $\omega_{pe} \cong \omega_p$.

Assuming the plasma is placed under the action of a sinusoidal electric field, $E_x = E_{ox}e^{j\omega t}$, the total current flowing through it will be:

$$j_{totx} = \varepsilon_o \frac{\partial E_x}{\partial t} + j_x \qquad (1.24)$$

where the two terms represent the displacement and the conduction current, respectively. Using (1.12), (1.20), (1.21) and (1.24), we obtain:

$$j_{totx} = j\omega\varepsilon_o \left[1 - \frac{\omega_p^2}{\omega(\omega - jv_c)} \right] E_x \qquad (1.25)$$

from which we can introduce the effective plasma dielectric constant:

$$\varepsilon_p = \varepsilon_o \left[1 - \frac{\omega_p^2}{\omega(\omega - jv_c)} \right] \qquad (1.26)$$

At atmospheric pressure the collision frequency is much larger than the driving frequency ($v_c \gg \omega$) and the equation (1.26) takes a simpler form:

$$\varepsilon_p \cong \varepsilon_o \left(1 - j\frac{\omega_p^2}{v_c\omega} \right) \qquad (1.27)$$

1.3.3

Electrical model of CCP

As an electrically conductive medium composed of negatively charged electrons of minimal mass, positively and negatively charged ions, electron excited metastable species plus radiative emission ($h\nu$), plasma state can be modeled as either a series or parallel resonant circuit. Each element within the circuit is a function of drive frequency of the electric field, energy coupling (resistive, capacitive, inductive or a combination of each) and plasma chamber reactance (due to construction materials and geometry).

One of the most widely used electrical equivalent models in the case of CCP is sketched in figure 1.3a, where C_o represents the plasma bulk capacitance, C_{sh} the total sheath capacitance, R_{bulk} the plasma bulk resistance and L_{bulk} the plasma bulk inductance.

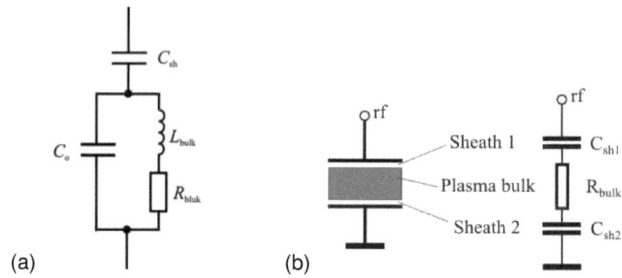

(a)　　　　　　　　　　　　　(b)

Figure 1.3 – Electrical models of CCP.

The plasma bulk inductance has two components: the geometrical inductance, L_g, and the electron inertia inductance, $L_e = R_{pl}/v_{en}$, where v_{en} is the electron–neutral collision frequency. The geometrical inductance is significant in inductively coupled plasmas but diminishes to near zero in CCP plasmas, and at atmospheric pressure, v_{en} is higher than 10^{12} s^{-1} so that the inductive reactance of the plasma is much lower than its resistance. Supposing that: (**a**) the ions respond only to time-averaging potentials; (**b**) the electrons respond to the instantaneous potentials and carry the RF discharge current; (**c**) the electron density is zero within the sheaths regions; (**d**) the ion density is uniform and constant in time everywhere in the plasma bulk and sheath regions, a simplified CCP generated between the discharge plates (electrodes) is shown in figure 1.3b. It consists of a series combination of plasma bulk resistance, R_{bulk} and of the sheaths capacitances C_{sh1} and C_{sh2}. The impedance of the plasma can be written as:

$$Z_{pl} = R_{bulk} + j\left(\omega L_{bulk} - \frac{1}{\omega \dfrac{C_{sh1}C_{sh2}}{C_{sh1}+C_{sh2}}} \right) \qquad (1.28)$$

Here:

$$C_{sh} = \varepsilon_o \varepsilon_{sh} \frac{A}{d_{sh}}$$ (1.29)

Where ε_o denotes the permittivity of vacuum, ε_{sh} denotes the dielectric constant of the plasma sheath, A is the common surface area of the electrodes and of the plasma (sheaths and bulk), and d_{sh} is the sheath thicknesses. The presence of the plasma causes both an increase in the capacitance and a loss resistance of the resonant circuit, consequently the resonance frequency reduces to a lower frequency and Q_u reduces to Q_L (figure 1.4).

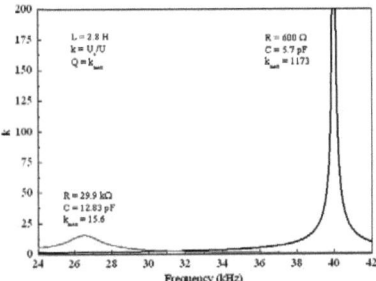

Figure 1.4 – Resonance curves for the same series circuit in the absence (right trace) and in the presence (left trace) of the plasma. [30]

This change of the resonance frequency is generally termed "frequency pulling". When the resonant LRC circuit is a component part of an oscillator, the frequencies of generated oscillations are different in the presence and in the absence of the plasma respectively. So, the "frequency pulling" can be a useful metrological tool.

2 Low power APP generators [2,3,5-8,30,50]

The material processing in atmospheric pressure plasmas has been given much attention during the last decades. Among atmospheric pressure plasmas, CCP is considered to be very useful in many applications mainly because of its relatively low breakdown voltage. Operating with or without dielectric layers between electrodes, CCP generators can produce non-equilibrium homogeneous glow like discharges having gas kinetic temperatures usually placed between 50 and 300 °C. Thus, CCPs can be thermally non-aggressive and are used for the treatment and processing of thermo-sensitive materials, especially polymers and other materials used in medicine, being able to generate chemically active species. A general recipe for obtaining non-thermal plasmas is the reduction of the discharge size and/or its duration. This effect is known from plasma display technology and from corona research, where streamer formation and transient character of the discharge may be seen as a natural size/time constriction.

The atmospheric pressure glow discharge can be generated in the capacitive mode by applying time variable electric fields (pulsed or sinusoidal) on different electrodes geometries. A comparative study regarding the efficiency of the two excitation waveforms achieved on an atmospheric pressure helium plasma jet has shown that the pulsed excitation is more advantageous for generating chemically active species. The pulsed excitation of the plan-parallel electrode configuration generates homogeneous discharges with relatively high electron number density, which can be used for the treatment of polymer surfaces.

Many of the commercial devices for generating atmospheric pressure non-thermal plasmas use generators having oscillating frequencies higher than 20 kHz. They are placed in the industrial ranges of 10-50 kHz, 4 MHz, 13.56 MHz, 27.12 MHz and 60 MHz. Sometimes, in order to optimize the transfer of the power from the generator to the plasma, a matching network is connected between the waveform generator and the plasma chamber. Very simple and low cost devices, capable of generating very low temperature atmospheric pressure plasmas, which might be of interest for the scientists who want to make such discharges, could be hand-made in their laboratories. The method is based on the possibility to generate very high voltages in *LRC* series circuits working under self-resonance conditions, as it was previously demonstrated.

In this section it will be shown how can be used the combination field effect transistor + resonant circuit in thermally non-aggressive plasma generators running at the frequencies lower than 2 MHz. In these devices the transistor is used as a switching element. It will be shown the experimental conditions under which very cold atmospheric pressure plasmas can be generated by electromagnetic fields with frequencies in the range of 0.04 – 1.6 MHz. The devices are supplied with low or very low DC voltages. The advantages which come from working at lower frequencies are: less pretentious technical requirements (especially the requirements regarding the impedances matching, the electromagnetic screening and the electrical insulation using special materials, and the requirements regarding the difficulties in measuring the high voltages and currents at very high frequencies), and the economic advantages arising from using a less expensive technology.

2.1

High voltage generators with resonant transformer

For generating high voltages able to ignite atmospheric pressure plasmas in different frequency ranges suitable self-resonant transformers must be used.

For lower frequencies (20 - 50 KHz) we have used a fly-back transformer of which the original primary coil was replaced by another one having only 6 turns. In order to generate non-thermal APP at frequencies slightly lower than 1 MHz a ferrite-core transformer with a transformation ratio 5/400 (primary/secondary) was constructed. The operating frequency can be increased to 1.5 - 2 MHz by replacing the ferrite-core transformer with a classical Tesla coil. In all situations the plasma chamber is connected to the free end of the transformer or of the Tesla coil, the other end being grounded. Figure 2.1 presents the schematic of the plasma generator together with the three variants of transformers. The mentioned frequencies are the self-resonance frequencies of the secondary circuits including the plasma chambers in different configurations.

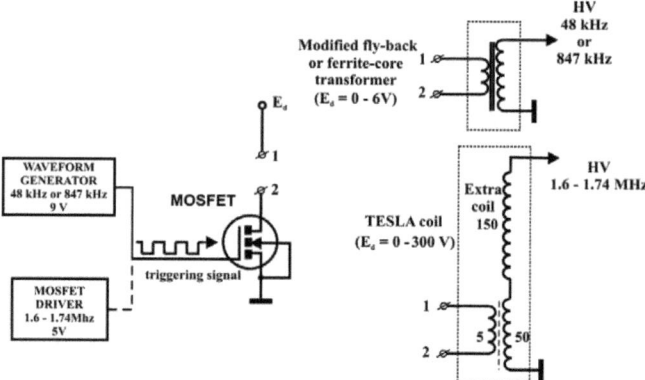

Figure 2.1 – Schematics of the high voltage generator for driving frequencies lower than 2 MHz. The resonance frequencies could have other values, depending on the plasma chambers geometry.

The primary coil of the transformer is connected in series with the channel of a field effect transistor. When one of the classical transformers (fly-back or ferrite core) is used, under the action of the gate voltage (the triggering signal) applied on its gate, the MOSFET acts as a switch between the primary coil and the ground and through the primary coil will flow current pulses having the same frequency as the gate voltage one. When the driving frequency is equal to the self-resonance frequency of the secondary circuit, the energy transfer from the primary coil to the secondary circuit will be maximum.

When the Tesla coil is used the working principle is the same, with the exception of the secondary circuit of the transformer. The Tesla coil (an air core transformer) has two stages of voltage increase. In the first stage the voltage is increased ten times by an air core transformer having 5 and 50 turns in primary and secondary windings, respectively. The second stage of the voltage increase is realized by adding a magnifier coil (150 turns) called "extra coil". The extra coil is not magnetically coupled with the air core transformer and its free end serves to supply the plasma chamber.

In all of the three configurations, depending on the plasma chamber geometry and on the plasma impedance, the transferred power from the generator to the plasma can be maximized by adjusting the frequency and eventually the duty cycle of the triggering signal.

2.2
Reactors for thermally non-aggressive APP

With the above described high voltage generators three kinds of thermally non-aggressive APP were generated: atmospheric pressure plasma micro jet (APPJ), dielectric barrier discharge (DBD) and plasma in contact with liquids (in or on liquid). For these aims, different plasma reactors were designed and constructed. It is worth mentioning that terms as "chamber", "container" or even "torch" instead of the term "reactor" are used to designate the device attached to the high voltage electrode for generating the plasma. In all situations one of the reactor's electrodes is connected to the free end of the secondary coil of the self-resonant transformer. It is called the powered electrode. The other one is usually grounded.

For generating AP very low power plasma jets, torch-like reactors having various dimensions were designed (figure 2.2). The device is composed of a metal wire electrode (1mm diameter) placed via a holder piece in a quartz tube with the inner diameter in the range of 0.8 – 4 mm through which plasma gas flows with adjustable flow rate. The electrode is made of kanthal (Fe 71.02%, Al-5.8%, Cr-22%, Mn-0.4%, Si-0.7% C-0.08%) and is connected to the generator output. The electrode is coated with glass or PTFE, except 1mm in length, to prevent the generation of corona discharge along it. The distance between the free end of the electrode and the exit of the quartz tube is adjustable (0.5 - 1 mm).

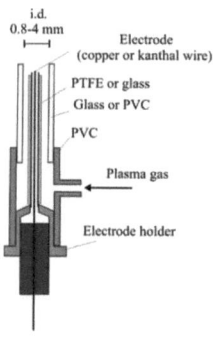

Figure 2.2 – Reactor for APPJ.

The chamber for generating an atmospheric pressure dielectric barrier discharge at 48 kHz excitation frequency is shown in figure 2.3 a. It consists of two horizontally positioned glass plates (0.8 mm thickness) vertically distanced at 3.1 mm by two glass spacers. The plates and spacers have the same width (30 mm). The horizontal distance between the spacers is of 10.3 mm. This parallelepipedic open ended box forms the plasma chamber. The electrodes (circular metallic plates, 16.5 mm diameter) are placed on the glass plates, outside and centred on the plasma chamber. By changing the value of DC supply voltage (E_d) from 1 to 6 V, the ranges of the electrodes voltage and of the current variation are 1.7 – 7.6 kV$_{pk-pk}$ and 0.6 – 5.7 mA$_{pk-pk}$, respectively. The working gas (helium, 0.44 l/min) is introduced through the back end of the chamber, it flows along the chamber perpendicularly to the electric field and exits into the free space through the other end of the chamber. Thus, the ambient air can penetrate into the plasma chamber by back diffusion.

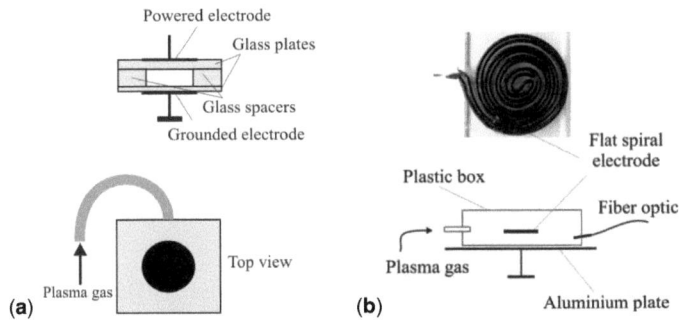

Figure 2.3 – Reactors for AP-DBD generated at 48 kHz.

Larger volumes of homogeneous DBD generated inside closed spaces with the excitation frequencies in the range of 20 - 50 kHz can be obtained with a plasma chamber like the one shown in figure 2.3 b. It is a commercial parallelepipedic transparent plastic box (90 x 45 x 34 mm^3) placed on a grounded aluminium plate. The powered electrode was made of a PVC coated copper wire which was bent in the middle to obtain a double conductor. It was used to make a flat spiral with four double turns which was placed between two glass plates (25 x 25 mm^2) and then the empty space was filled with melted plastic. The two free ends of the double conductor were welded and connected to the hot end of the secondary coil of the fly-back transformer, where a sinusoidal voltage of 5 kV$_{pk-pk}$ and 48 kHz is generated. The plasma can be ignited in flowing helium (the plastic box is not hermetically closed) with DC powers of few watts. The DC drain voltage for supplying the circuit is very low (4 V in our experiment).

23

For generating an AP-DBD at 1.74 MHz a parallelepipedic plasma chamber with a total volume of 160 cm^3 was designed (figure 2.4). It contains two disk shaped metallic electrodes (24.5 mm diameter) covered with PTFE (1.5 mm thickness) as dielectric. The gap of the discharge space can be modified in the range of 0.5-2 cm. One of the electrodes is connected to the output of the drive generator and the other one is grounded. The chamber has an access nozzle through which helium gas with flow-rates lower than 5 l min^{-1} is introduced. It flows perpendicularly to the electric field through the discharge space and is exhausted through two exit holes placed on the opposite side of the chamber. In the same time, small quantities of ambient air can penetrate by back-diffusion in the plasma chamber through the two exit holes.

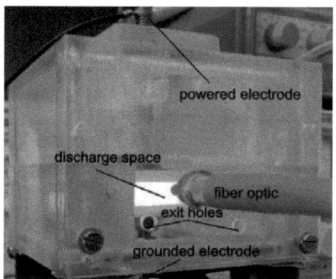

Figure 2.4 – Reactor for AP-DBD generated at 1.74 MHz. [50]

When a device like the one shown in figure 2.2 is vertically immersed inside a liquid and the plasma gas is bubbled, a cold gas discharge can be generated in liquid. Figure 2.5 shows a sketch of this experimental arrangement and the associated plasma image. When the plasma gas (usually argon or helium) flows, at the immersed exit of the quartz tube gas bubbles arise with a cadency of 0.5 – 10 bubbles/second, depending on the gas flow rate (max. 0.3 l min^{-1}). During each bubble life time, between the free end of the electrode and the bubble wall a glow discharge is generated, the bubble wall representing the secondary electrode with a floating electric potential. Alternatively, when the torch-like device is vertically placed over the surface of the liquid, for a gas flow-rate of 0.5 l.min^{-1} a stable and relatively uniform discharge can be generated for an electrode-liquid distance of maximum 10 mm. As it can be observed, in our experiment a single metal electrode is used for generating the plasma in or on liquid. This represents an important advantage compared to other generation systems where, to generate APPin contact with liquids, two electrodes (a powered one and a grounded one) are used. This is possible by choosing the most appropriate geometric dimensions of the device in correlation with the optimum gas flow-rate and electrical parameters.

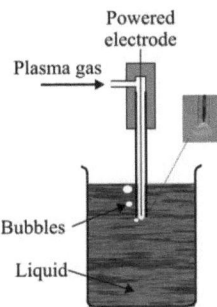

Figure 2.5 – Sketch and a detailed image of in-liquid plasma reactor.

3 Electrical analysis [2,3,5]

The measurement of the voltage, current and power of atmospheric pressure plasmas is not a trivial task. There are two main methods used in this purpose. The first method can be used when the plasma reactor is directly connected to the AC power supply and it is based on the direct measurement of the voltage on the plasma reactor and of the current flowing through the plasma. This is an invasive method because it uses voltage and current probes which can perturb the plasma and can induce an additional phase angle between the current and the voltage. The second method is used when the plasma reactor is connected to the output of the AC power supply via a matching network which has the role of providing the maximum energetic transfer from power supply to the plasma. Under matching conditions, in the presence of a discharge inside reactor, the reactive part of the plasma impedance is compensated by the reactive part of the matching network and the whole circuit will have a resistive behavior. The plasma power can be now measured with a power-meter and the voltage across the plasma reactor is measured with a very high impedance voltage probe. Knowing the matching network characteristics, the plasma power and the voltage across plasma reactor can be calculated the resistance and the reactance of the plasma.

In this section other noninvasive methods for determining the electrical plasma parameters are shown. They are based on electrical models of plasmas. Based on the electrical analysis of the circuits composed of the high voltage generator and plasma reactor in the presence and in the absence of plasma we can have important information on the plasma characteristics, such as power density, current density, electron number density or plasma developing stages. Depending on the working frequency of the electric field we have used two computing ways. The first one is applied at higher frequencies when the direct measurement of the voltage on the powered electrode is almost impossible because of the influence of the high voltage probe impedance on the self-resonant circuit characteristics. The second one is based on the direct measurement of the powered electrode voltage and of the current flowing through the plasma. Both methods are associated with electrical models of the plasma reactor for obtaining information on plasma characteristics.

3.1

Electrical analysis of plasma micro-jet

An atmospheric pressure plasma micro-jet can be generated by using a combination high voltage generator + ferrite-core transformer (figure 2.1) + torch-like reactor (figure 2.2). The transformer is a high voltage one, the primary winding having 5 turns and the secondary 400 turns. The two windings are placed side-by-side on o ferrite rod (10 mm diameter). This kind of windings placement prevents the breakdown between turns or between coils. When the plasma reactor (torch) is supplied with helium, argon or their mixture a jet-like discharge can be generated at the end of the powered electrode. The plasma is not in direct electrical contact with a second electrode.

Before the analysis of the circuit it is important to know how it works. In the circuit consisting of DC power supply, MOSFET channel and the primary coil of the transformer (with the inductance L_1 and the loss resistance r_1') the last component represents the load impedance. The MOSFET acts as an electronic switch driven by the gate triggering signal which has a square waveform (unipolar, 10V) with adjustable frequency and duty cycle. When the triggering signal is in the high state, the MOSFET channel is in the low resistance state and through the primary coil are flowing current pulses with the same frequency as that of the triggering signal.

The secondary coil of the transformer is magnetically coupled with the primary coil. The induced voltage from the primary coil to the secondary coil represents the supply voltage for the secondary circuit consisting of secondary coil + plasma chamber (in the absence of the discharge) or of secondary coil + plasma chamber + plasma (in the presence of the discharge). The secondary coil of the transformer has many turns (it is a high voltage transformer) and its self-capacitance may not be ignored in the electrical analysis of the circuit. The secondary coil of the transformer can be modeled as a circuit consisting of the inductance L_2, the self-capacitance C_2 and the loss resistance r_2. When the plasma reactor (chamber) is connected in the secondary circuit it adds a capacitance (C_T) and a loss resistance (R_T). The equivalent circuit in this situation is shown in figure 3.1. The loss resistance of the plasma chamber is much lower than the loss resistance of the secondary coil and its contribution to the total resistance of the circuit can be ignored in the circuit analysis.

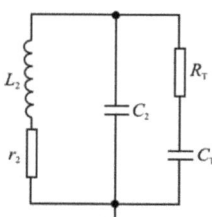

Figure 3.1 – Equivalent circuit of the secondary circuit in the absence of the plasma. [3]

With these considerations in mind, the equivalent circuit of the high voltage transformer is shown in figure 3.2a where: E_1 includes all contribution to the electric force (DC power supply and sinusoidal voltage reflected by electromagnetic induction from the secondary coil to primary coil), r_1 includes the loss resistance of the primary coil and the resistance of the MOSFET channel, M is the mutual inductance of the transformer and $C_{2e} = C_2 + C_T$.

Because the capacitance of the primary coil is very small and it works at a frequency much smaller than the self-resonance frequency of the primary circuit, its effect was ignored. The values of capacitances C_2 and C_T can be determined from the resonance frequencies of the secondary circuit in the presence and in the absence of the plasma reactor, respectively.

When the discharge is a plasma jet it has a purely resistive behavior with the resistance R_{pl} and the equivalent circuit is presented in figure 3.2b.

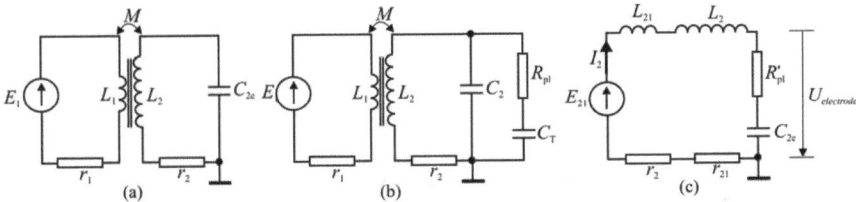

Figure 3.2 – Equivalent schemes: (**a**) of the whole circuit in the absence of the plasma; (**b**) of the whole circuit in the presence of plasma; and (c) of the magnetically coupled secondary circuit in the presence of the plasma. [3]

The primary circuit is supplied with square current pulses composed from many sinusoidal signals (Fourier components), the main component having the frequency equal to the frequency of current pulses. The secondary circuit, magnetically coupled with the secondary one, is an *LRC* series circuit which acts as a band-pass filter (selective filter) and from all sinusoidal components it selects only that component having the frequency equal to its natural frequency. The components with different frequencies are drastically attenuated. The secondary circuit being a series one, the voltage between the ends of the secondary coil and implicitly between the powered electrode of the plasma reactor and the ground will be maximum and an AP electrical discharge can be generated. The power transfer from the primary circuit to the secondary circuit and implicitly to the plasma will be maximum when the frequency of the triggering signal (f_{exc}) is equal to the self-resonance frequency of the secondary circuit (f_o).

Figure 3.3 shows the voltage waveforms in different points of the diagram when it works under the maximum energy transfer to the plasma. The voltage between the ends of the secondary coil was monitored by using an induction method and the voltage values in plot D are arbitrary.

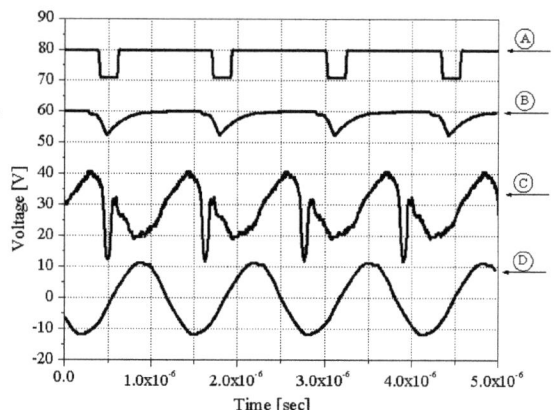

Figure 3.3 – Signal waveforms: (**A**) – the waveform of the voltage excitation pulses registered when the pulse generator was not connected to the gate of MOSFET; (**B**) – the real waveform of the voltage excitation pulses when the pulse generator was connected to the gate of the MOSFET; (**C**) – the waveform of the voltage between the ends of the primary coil of the transformer determined by the current pulses and by the reflected voltage from the secondary circuit. (**D**) – the sinusoidal voltage between the ends of the secondary coil. To avoid the superposition of the traces, the plots A, B and C were shifted with 80, 60 and 30V with respect to voltage axis. [2]

The plasma can be also generated at the driving frequencies lower than the optimum one but only if $kf_{exc} = f_o$ (k – integer) and with a lower efficiency transfer of the power to the plasma. This is because between two excitation pulses the amplitude of the sinusoidal voltage in the secondary circuit decreases exponentially as it can be seen in figure 3.4a for $k = 4$. Figure 3.4b shows the optimum operating situation, when $f_{exc} = f_o$.

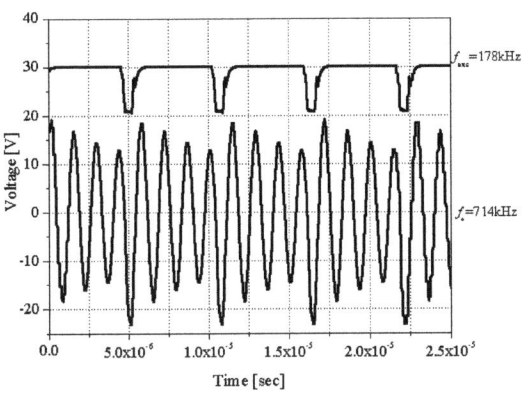

(a)

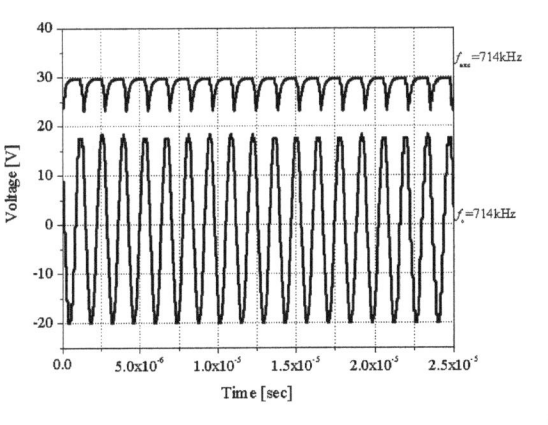

(b)

Figure 3.4 – Waveforms of the excitation pulses and of the voltage between the ends of the secondary coil at two different excitation frequencies: (**a**) $f_{exc} = f_0/4$, and (**b**) $f_{exc} = f_0$. The plots of the excitation signal were shifted with 30V with respect to voltage axis. [2]

Because of high frequency and of high voltage in the secondary circuit, the direct measurement with voltage and current probes is very difficult to do. The presence of an electric probe determines the mismatch of the secondary circuit and incorrect results of measurements. Moreover, the measurement results can also be influenced by the electromagnetic radiations of the

31

generator. Therefore, in order to find the plasma characteristics we have adopted the computing way which requires knowing the electrical information only from the primary circuit (the measured voltages on the two ends of the primary coil of the transformer and the mean current flowing through it) and the transformer characteristics (inductances, loss resistances and capacitances of windings) which can be determined by bridge methods.

The primary coil of the transformer is under the action of two distinct voltages: a sinusoidal one reflected from the secondary circuit and a squared one determined by the current pulses injected by the DC power supply due to the switching behavior of the MOSFET. They have the same frequency as the self-resonance frequency of the secondary circuit and implicitly of the triggering signal. Therefore the primary circuit consisting of primary coil connected in series with the MOSFET channel can be analyzed in two ways: by studying its behavior in the sinusoidal (harmonic) regime and in the switching regime, respectively.

The analysis in the sinusoidal regime is done at the self-resonance frequency of the secondary circuit which represents also the working frequency. The secondary circuit can be represented as a simpler series circuit by using the transfiguration dipole theorem and the harmonic analysis. It shown in figure 3.2c, where:

$$R'_{pl} = R_{pl}\left(1+\frac{C_2}{C_T}\right)^{-2} \tag{3.1}$$

$$E_{21} = \frac{\omega M}{r_1^2 + \omega^2 L_1^2} E_1 \tag{3.2}$$

$$L_{21} = -\frac{\omega^2 M^2}{r_1^2 + \omega^2 L_1^2} L_1 \tag{3.3}$$

$$r_{21} = \frac{\omega^2 M^2}{r_1^2 + \omega^2 L_1^2} r_1 \tag{3.4}$$

Electrical analysis

The resonance frequency of the inductively coupled secondary circuit will be:

$$f_o = \frac{1}{2\pi\sqrt{(L_2 + L_{21})C_{2e}}}$$ (3.5)

In the previous equations the index "21" refers to the influence of the primary circuit over the secondary circuit via magnetic coupling with the mutual inductance M. This influence is materialized by an increase of the total loss resistance and a decrease of the total inductance of the secondary circuit. Working at the self-resonance frequency of the secondary circuit the voltages on its reactive elements, L_2+L_{21} and C_{2e}, will be maximum. This means that the voltage between the free end of the plasma electrode and the ground will be maximum.

In the sinusoidal regime, the influence of the secondary circuit on the characteristics of the primary circuit depends on the magnetic coupling via $\omega M/Z_2$ factor, where Z_2 represents the impedance of the secondary circuit in the absence of the plasma. Having in view that Z_2 is very high and ωM is very small, the influence of the secondary circuit on the reactance of the primary circuit can be ignored. During the experiment two things were observed. Firstly, a sinusoidal voltage with the frequency f_o is induced in the primary coil. It has few volts and the circuit analysis shows that:

$$E_{12} = \frac{\omega M}{Z_2} E_{21}$$ (3.6)

It can be easily measured with two voltage probes and then can be used to calculate the induced sinusoidal voltage E_{21}.

Secondly, when the plasma is generated in the secondary circuit the current absorbed by the whole circuit from the DC power supply decreases. This is because the total loss resistance of the primary circuit increases with

$$\Delta r_1 = \frac{\omega^2 M^2}{Z_2^2} R_{pl}'$$ (3.7)

33

what represents the reflected resistance from the secondary circuit. It can be determined by finding the relationship between the current consumption in the absence and the presence of the plasma and the total resistance of the primary circuit in the two situations. This relationship is found following the Laplace formalism for analysis of circuits working in switching regime (see *Appendix*). In the equivalent Laplace scheme of the primary circuit (figure 3.5) r_1 represents the total resistance of the primary circuit consisting of the loss resistance of the primary coil, the resistance of the MOSFET channel and the resistance reflected from the secondary circuit. The switch K simulates the state of the channel, very high or very low resistance, depending on the level of the triggering signal. When the triggering voltage is high the channel resistance is very low (K is closed) and when the triggering voltage is low the channel resistance is very high (K is opened).

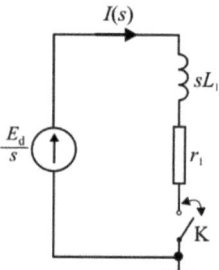

Figure 3.5 – Laplace model of the primary circuit. [3]

When the circuit is closed, the Laplace current is expressed as follows:

$$I_1(s) = \frac{E_d}{L_1} \cdot \frac{1}{s\left(s + \frac{L_1}{r_1}\right)} \qquad (3.8)$$

and the corresponding current in the time domain is:

$$i_1(t) = \frac{E_d}{r_1} \cdot \left(1 - \exp^{-\frac{r_1}{L_1}t}\right) \qquad (3.9)$$

The mean intensity of the current flowing through the primary coil can be directly measured with an ammeter. Its theoretical expression is:

$$\langle i_1 \rangle = \frac{E_d}{r_1 T} \cdot \left[t_1 + \frac{L_1}{r_1} \left(\exp^{-\frac{r_1}{L_1} t_1} - 1 \right) \right]$$ (3.10)

where T and t_1 can be determined from the waveform of the triggering signal. It is used to calculate the resistances of the primary circuit in the absence and in the presence of the plasma in the secondary circuit and of the difference Δr_1 between them. Knowing the resistances of the primary circuit and using the equations (3.2), (3.3), (3.5) and (3.7) one can calculate E_{21}, L_{21}, r_{21} R'_{pl} and R_{pl}. Based on the equivalent scheme of the secondary circuit (figure 3.4c) which has a resistive behavior (it works under resonance state) can be calculated the current flowing through this circuit and the voltage between the powered electrode and ground:

$$I_2 = \frac{E_{21}}{r_1 + r_{21} + R'_{pl}}$$ (3.11)

$$U_{electrode} = I_2 \sqrt{R'^2_{pl} + \frac{1}{\omega^2 C^2_{2e}}}$$ (3.12)

Finally, plasma electrical characteristics (the intensity of the current flowing through the plasma chamber and the power consumed by the plasma) are calculated based on the equivalence of secondary circuits shown figures 3.4b and 3.4c:

$$I_{pl} = \frac{U_{electrode}}{\sqrt{R^2_{pl} + \frac{1}{\omega^2 C^2_T}}}$$ (3.13)

$$P_{pl} = I^2_{pl} R_{pl}$$ (3.14)

The analysis method shown in this section seems to be very complicated but it has the advantage of using the results of direct measurements, done exclusively in the primary circuit of the transformer without perturbing the secondary circuit. We should not forget that in the secondary circuit the plasma is generated and it runs under resonance state, being very sensitive to the action of the perturbing factors. Summarizing, the development of this method is based on two kinds of parameters: parameters which can be directly measured or determined by measurements (the transformer characteristics L_1, L_2, r_2 and M, the frequency and the duty cycle of the triggering signal, the voltage of the DC power supply (E_d) and the mean intensity ($<i_1>$) of the absorbed current from the DC power supply) and parameters (plasma characteristics) which are calculated based on the theoretical analysis of the circuit.

3.2
Electrical analysis of dielectric barrier discharge

Atmospheric pressure dielectric barrier discharge can be generated by using a combination high voltage generator + ferrite-core or Tesla transformer (figure 2.1) + DBD reactor (figure 2.3 or 2.4). In our computing example the situation in which a combination of a modified fly-back transformer and a glass plasma container (figure 2.3 a) is analyzed. The working frequency is of 48 kHz.

In technological applications when APDBD is used for surface treatment, the plasma must be in its homogeneous developing stage characterized by uniform distribution of light emission. In this stage the plasma can be modeled as a sandwich structure, as it is shown in figure 1.3b. Having in view that in the case of DBD the two electrodes are covered with dielectric layers, the plasma container can be sketched as in figure 3.6a. Supposing (for simplicity) that the two sheaths and the two dielectric layers have the same electrical and geometrical characteristics, the plasma container can be modeled as in figure 3.6b, where:

$$C_{tot} = \frac{C_{sh}C_d}{2(C_{sh}+C_d)} \qquad (3.15)$$

Here, C_{sh} and C_d are the series equivalent capacitances of the two pairs of sheath and dielectric capacitances, respectively.

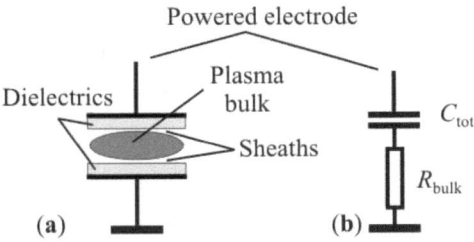

Powered electrode

Dielectrics

Plasma bulk

Sheaths

C_{tot}

R_{bulk}

(a)   (b)

Figure 3.6 – Electrical model of DBD container.

When the RF voltage applied on the powered electrode and the intensity of the current flowing through the plasma sandwich structure are known (measured or determined) the next group of equations can be used for estimating the DBD characteristics:

$$Z_{tot} = \frac{V}{I} \qquad (3.16)$$

$$Z_{tot}^2 = R_{bulk}^2 + \frac{4(C_{sh}+C_d)^2}{\omega^2 C_{sh}^2 C_d^2} \qquad (3.17)$$

$$R_{bulk} = \sqrt{\frac{V^2}{I^2} + \frac{4(C_{sh}+C_d)^2}{\omega^2 C_{sh}^2 C_d^2}} \qquad (3.18)$$

$$p = \frac{I^2 R_{bulk}}{plasma\ volume} \qquad (3.19)$$

$$j_e = \frac{I}{A} \qquad (3.20)$$

$$n_e = \frac{j_e}{e\mu_e \frac{IR_{bulk}}{d_{bulk}}} \qquad (3.21)$$

In the writing of the above equations the applied electric field and the current flowing through the circuit were approximated as having a sinusoidal time variation with the angular frequency ω. The capacitances of sheaths and dielectrics were calculated based on formulas like that in the section one

(1.29). The dielectric constant of the plasma can be taken from literature (books or journal papers). For our example (AP helium DBD, 48 kHz, see the next section) ε_{sh} = 1.52 and the geometrical characteristics were: A = 163 mm^2, d_{sh} = 0.15 mm, d_d = 0.8 mm and *plasma volume* = 0.5 cm^3. The dielectric material was glass with ε_d = 1.52. The electron mobility in helium at atmospheric pressure is 1128 cm^2/(Vs). The contribution of the positive helium ions to the total current was ignored because their mobility is much lower than the electron mobility.

4 Discharge characteristics [2,3,5-8,30]

4.1
Micro-jet characteristics

This section presents the results of the study done on an AP plasma jet generated at 714 KHz by using the combination high voltage generator + ferrite-core transformer + plasma reactor for APPJ (figures 2.1 and 2.2). This working frequency is the self-resonance frequency of the secondary circuit in the presence of the plasma. The values of the electrical characteristics of the discharge were calculated based on the algorithm developed in the section 4.1. The plasma gas is helium whose flow-rate was monitored with a mass flow controller. Plasma characteristics depend both on the voltage (E_d) of the DC power supply and on the helium flow-rate.

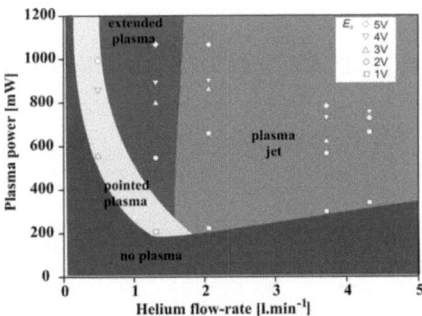

Figure 4.1 – Plasma power as function of helium flow-rate and the three development stages of the plasma. [3]

Figure 4.1 presents the dependence of the plasma power on the helium flow-rate for different DC voltages and the delimitation of areas corresponding to different developing stages of the discharge. Additionally, in figure 4.2 are shown the images of the plasma in its developing stages. Some conclusions can be drawn from the obtained results and from visual observation.

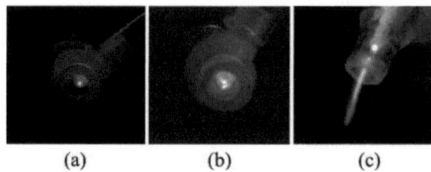

(a)   (b)   (c)

Figure 4.2 – The three developing stages of the helium plasma jet: (**a**) pointed plasma; (**b**) extended plasma; and (**c**) plasma jet. [3]

Firstly, the back-diffusion of air in the plasma has a high influence on the minimum DC voltage needed for plasma generation. This is because the breakdown voltage of air is higher than that for helium. The influence of back-diffusion of ambient air can be reduced by the increase of plasma gas flow-rate. This is the reason why at lower DC voltages (1-2 V) the plasma can be generated only for helium low-rates lower than 1.3 l min^{-1}. When the DC voltage is higher than 2 V and the back diffusion of ambient air is lower, the plasma can be also generated at helium flow-rates lower than 1.3 l min^{-1}.

Secondly, it can be observed that for a constant flow-rate of helium the plasma power increases with the increase of the DC voltage. This is a normal behavior because the voltage on the powered electrode of the plasma reactor increases with the increase of DC voltage.

Thirdly, it can be also observed that, with the exception of $E_d = 1$ V, for a constant DC voltage the plasma power presents a maximum. The position of this maximum depends on the helium flow-rate and it is located at a lower flow-rate for a higher DC voltage. Its presence can be explained by the dependency of the energy transferred from the plasma to the ambient air on the helium flow-rate and implicitly on the helium flowing velocity. It increases at higher flow-rates and the energy retained by the plasma decreases. The position of the maximum power can also be correlated with the spatial developing stages of the plasma for different helium flow-rates: pointed plasma, extended plasma and plasma jet (figure 4.2). The maximum absorbed power by the plasma is associated with the transition from the extended stage to the jet stage. When this transition takes place, the plasma length and the volume increase significantly, the energy transferred to the ambient air also increases and the plasma power becomes lower.

The calculated voltages on the powered electrode in the absence of the plasma were between 1.03 kV for E_d = 1 V and 1.88 kV for E_d = 5 V. In the presence of the plasma jet it increases with 350V and 650 V, respectively. Depending on helium flow-rate the calculated current flowing through the plasma is in the range of 1-2.5 mA. It was also measured with a commercial current probe. Figure 4.3 shows only a small difference between the two sets of values, confirming that our developed computing algorithm can be successfully used in the analysis of such plasma generators. We have to mention that the use of the current probe was only a way to validate the computing algorithm because it can disturb the optimum working parameters of the generator and some corrections (such as the readjustment of the driving frequency) must be done.

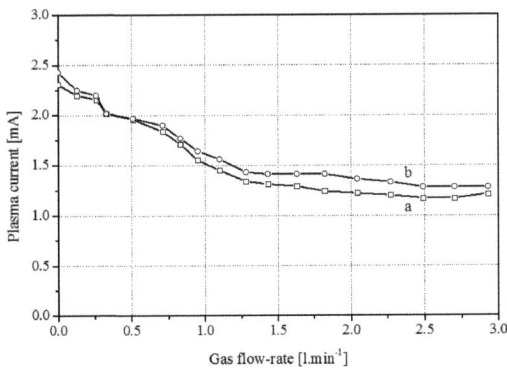

Figure 4.3 – (**a**) calculated, and (**b**) measured, current intensity flowing through the plasma as function of gas flow-rate, for E_d = 3 V and an axial distance of 0.5 mm. [3]

When the plasma is in the jet developing stage the efficiency of the energetic transfer from the DC power supply to the plasma is in the range of 35-56 %, depending on the DC voltage and on the helium flow-rate.

The axial temperature at a distance of 1.5 mm from the powered electrode was measured with a commercial thermocouple of whose junction was covered with a glass cap. Thus, the metal junction does not disturb the plasma. Figure 4.4 presents the dependence of the helium plasma jet on the helium flow-rate for a DC voltage of 3 V. As it can be seen the plasma temperature is slightly higher than the temperature of ambient air and it presents a maximum for helium flow rates in the range of 0.9-1.4 l min^{-1}. This maximum corresponds to the maximum power absorbed by the plasma.

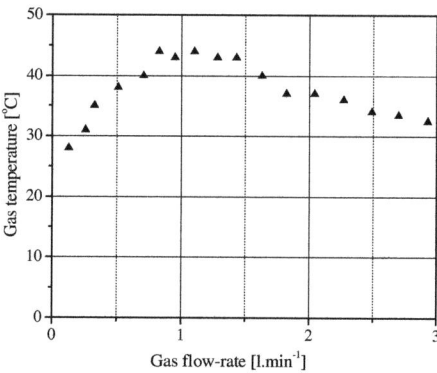

Figure 4.4 – Dependence of plasma temperature on the He flow-rate at an axial distance of 1.5 mm on the plasma electrode tip. DC supply voltage, E_d = 3 V. [2]

The thermally non-aggressive character of the helium plasma jet is confirmed in the image from figure 4.5. It can be controlled by the correct adjustment of the working parameters: the voltage of the DC power supply, the helium flow-rate and the distance from the powered electrode to the object exposed to the plasma action.

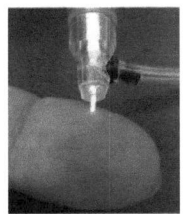

Figure 4.5 – Non-thermal plasma. He flow rate: 2.35 l/min, plasma power 680 mW. [2]

Plasma jets like the one analyzed in this section can also be generated in argon or air or in mixtures of helium, argon and air. Figure 4.6 supports this affirmation but it has to be mentioned that argon and air plasma jets are thermally aggressive, their temperatures being higher than 100 °C (110 °C and 137 °C, respectively). Moreover, argon and air plasma jets are less stable and have ramification tendencies.

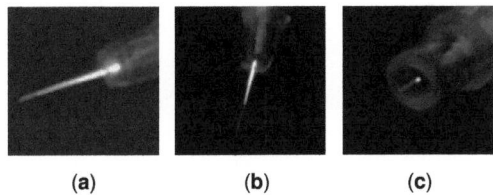

(a)   (b)   (c)

Figure 4.6 – Plasmas at atmospheric pressure: (**a**) He 3 l/min, 0.96 W; (**b**) Ar 2.5 l/min, 1.16 W and (**c**) open air, 1.44W. DC supply voltage, E_d = 5V. [2]

By replacing the ferrite-core transformer with the modified fly-back transformer the working frequency is reduced to 48 kHz. Figure 4.7 shows the helium plasma jet appearance and the plasma temperature under different operating conditions. The helium flow rate and the DC supply voltage (E_d) were kept constant while the inner diameter of the quartz tube of the plasma reactor was modified. Figure 4.7a shows that the length of the plasma column, the plasma power and the plasma temperature can be controlled by adjusting the quartz tube diameter. The axial gas temperature was measured with the glass-covered thermocouple without disturbing the plasma.

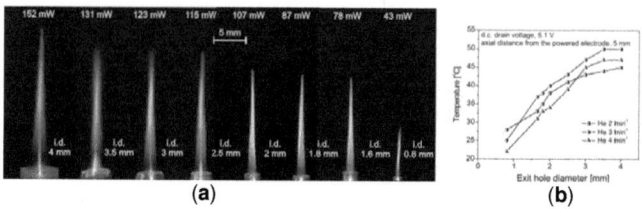

Figure 4.7 – (**a**) 48 kHz plasma jet appearance and plasma power as function of inner diameter of the exit nozzle of the glass tube (He flow rate, 4 l min⁻¹), and (**b**) gas temperature for different helium flow-rates. [8]

Based on the analysis of the plasma emission spectra, thermal characteristics of plasma jets and electron number density were evaluated (Table 4.1). The electron excitation temperature of helium atoms was estimated based on the Boltzmann plot method, the temperatures of excitation of rotational states of the hydroxyl radicals and of the nitrogen ionic molecules were estimated by finding the best fit (chi-square method) of the measured molecular spectra with the synthetic spectra generated by a dedicated software and the electron number density was calculated based on the Stark broadening of hydrogen emission line, H_α. By comparing the obtained results it can be concluded that the helium plasma jet generated at 48 kHz is more suitable to be used as thermally non-aggressive plasma. It is colder and the electron excitation temperature is very close to that of 714 kHz plasma jet which means that it can be a suitable source of chemically active species for materials processing.

Table 4.1. APP helium plasma jet characteristics.

f [kHz]	Q_{He} [l·min⁻¹]	P [mW]	T_{excHe} [K]	T_{rotOH} [K]	$T_{rotN_2^+}$ [K]	n_e [cm⁻³]
48	3	152	2340	360	392	3.69 10¹²
714	1.0	510	2730	470	540	1.03 10¹³

4.2

DBD characteristics

It is well known that the atmospheric pressure gas discharges are inherently susceptible to the glow-to-arc transition, affecting their stability and homogeneity and increasing the gas temperature. It has been demonstrated that dielectric barriers placed between two planar or cylindrical AC powered electrodes provide an effective control over unlimited current growth and allow atmospheric pressure glow discharges to be operated at very high current densities without being endangered by the glow-to-arc transition. By covering one or the both electrodes of the plasma chamber with dielectric materials (glass, quartz, ceramics, or polymer-materials with low dielectric loss and high breakdown strength) the homogeneity, temperature and plasma activity can be better controlled.

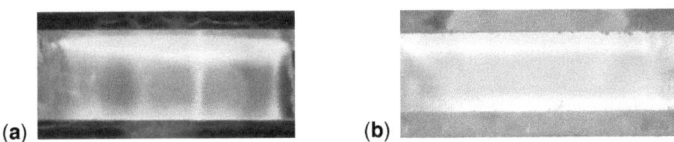

(a)　　　　　　　　　　　　　　(b)

Figure 4.8 – Filamentary (**a**) and diffuse (**b**) AP helium DBD at 48 kHz.

Dielectric barrier discharges have two developing stages: the non-homogeneous (filamentary) and homogeneous (diffuse) stage, respectively (figure 4.8). The non-homogeneous stage is characterized by a non-uniform distribution of light emission and by the presence of more than one major peak per half period of the current flowing through the plasma. The homogeneous stage is characterized by a uniform distribution of light emission and by the presence of a single major current peak. The generation of DBD having a diffuse or filamentary character can be controlled by a suitable adjustment of working parameters: the driving frequency of the electric field, the plasma power and the kind and the flow-rate of major plasma gas. The presence of metastables atoms can also be important.

In an earlier section we have shown that by attaching a glass plasma chamber, such as that shown in figure 2.3a, to a high voltage generator with a modified fly-back transformer (figure 2.1) an atmospheric pressure helium dielectric barrier discharge at 48 kHz can be generated. The ambient air can penetrate into the plasma chamber by back diffusion.

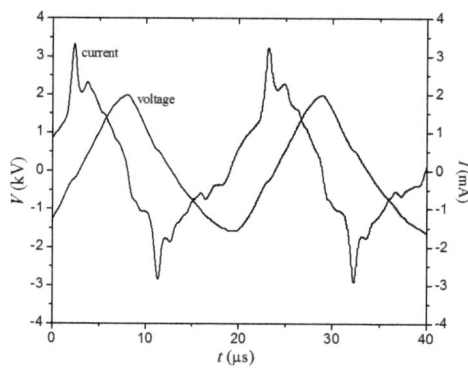

Figure 4.9 – Electrode voltage and current waveforms of AP homogeneous DBD generated at 48 kHz.

The characteristic parameters of plasma can be controlled by adjusting the DC supply voltage, E_d. In our experiment, in order to generate a homogeneous AP DBD the DC supply voltage has to be in the range of 1.5 – 6 V. The homogeneous or non-homogeneous (filamentary) character of the plasma was monitored via voltage and current waveforms visualization (figure 4.9), knowing that the presence of a single major current peak during a half period of the electric field characterizes a DBD in its homogeneous developing stage. Moreover, the phase angle between current and voltage (the current leads the voltage) confirms the capacitive character of the discharge.

The volt-ampere characteristic of AP helium DBD (figure 4.10) corresponds to the two specific discharge regimes, normal and abnormal, characterized by the relative constant voltage and by ascending current, respectively. When the current becomes higher than 2 mA the discharge transits from homogeneous stage to filamentary stage.

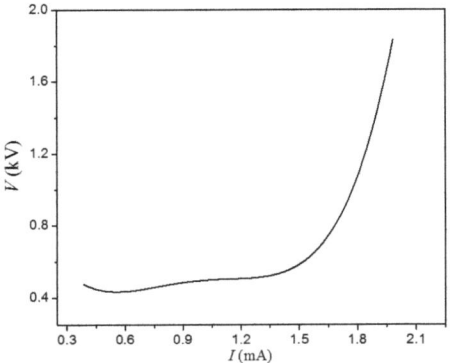

Figure 4.10 – Voltage-current characteristic of AP helium DBD at 48 kHz.

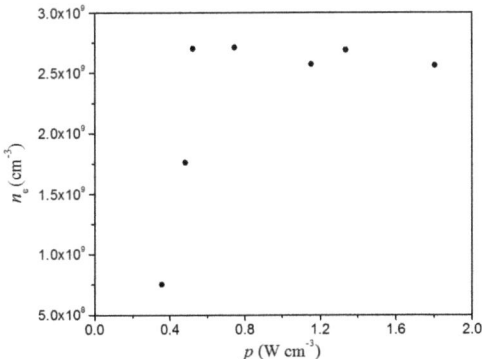

Figure 4.11 – Electron number density dependence on the plasma power density.

The dependence of the electron number density on the plasma power density is shown in figure 4.11. In the homogeneous developing stage, the saturation tendency is evident. When the power density exceeds 2 W cm^{-3} the plasma becomes filamentary and the results obtained with the considered electric model are inconsistent. The model was developed only for homogeneous developing stage of the plasma. Electron number densities of

$10^9 - 10^{10}$ cm^{-3} are characteristic for AP helium DBD with power densities lower than 2 W cm^{-3}.

The characteristic parameters of AP helium DBD generated with an electric field having a frequency of 48 kHz calculated according to the algorithm developed in the section 3.2 are presented in Table 4.2

Table 4.2. AP helium DBD characteristics.

p [W cm^{-3}]	j_e [mA cm^{-2}]	T_{gas} [K]	n_e [cm^{-3}]
2	1.8	323	2.5 10^9

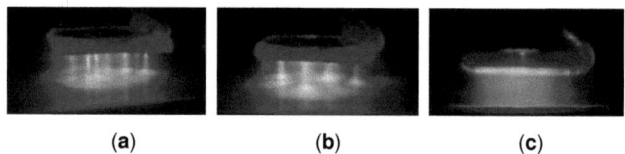

(a)   (b)   (c)

Figure 4.12 – Three plasma developing stages: **(a)** filaments – 0.2 l/min He, 1.8 W; **(b)** columns – 1.4 l/min He, 2.1 W; **(c)** homogeneous – 2.1 l/min He, 2.3 W. [7]

When a plasma chamber is used, as in figure 2.3b larger volumes of homogeneous DBD can be generated. Depending on the helium flow-rate, the plasma has three main developing stages. For helium flow-rates in the range of 0.2 – 0.6 l/min the plasma consists of several thin and unstable bright filaments (figure 4.12a). Their diameters are of around 1 mm. Our supposition is that the higher number of filaments than when a compact flat electrode is used is determined by the particular construction of the flat spiral electrode: more coplanar electrically isolated conductors having the same potential. The filaments have a self-organizing tendency, so that their distribution along the equipotential conductor is relatively uniform, which will favor the further developing of a homogeneous discharge covering the whole surface of the electrode.

With the increase in the helium flow-rate, the filaments diameters increase, their number decreases and the discharge becomes more stable. Around the value of 0.6 l/min helium flow-rate, the discharge is very stable and consists of seven diffuse columns with diameters of 2-3 mm (figure 4.12b). The discharge columns are located on a circle, being relatively equidistant. At the contact with the bottom wall of the plasma chamber (which is in contact with the outer grounded metallic plate) each column has an independent superficial luminescence. The further increase in the helium flow-rate up to 1.8 l/min is accompanied by the increase in the column diameters until reaching the third evolution stage when the plasma has a homogeneous appearance (figure 4.12c). The plasma column is diffuse, uniform and less luminous. At the contact with the bottom wall of the plasma chamber, the plasma spreads on the plastic surface forming a uniform luminous area with a diameter of 35 mm. This is important for the surface treatment of various materials placed in the plasma chamber. During the evolution from the filamentary stage to the homogeneous stage, the plasma power increases slightly from 1.8 W to 2.3 W. The plasma power was estimated simply by taking the difference between the consumed powers from DC voltage source in the presence and in the absence of the discharge respectively.

In the homogeneous developing stage, the measured gas kinetic temperature was of 51 °C and the estimated mean plasma power density was of 370 mW/cm^3.

An atmospheric pressure dielectric barrier discharge can also be generated at higher working frequencies, in the MHz range for example. The main particularities of the plasma generator (1.74 MHz, helium) are the use of a classical Tesla coil (instead of modified fly-back transformer) and that of a higher DC power supply which feeds the circuit (300 V as against 5-10 V). Its working principle was presented in the section 2.1.

The electrical diagnostic of the plasma is based on the electrodes voltage determination and on the measurement of the phase angle between the voltage across the plasma chamber and the current which flows through it. The waveforms of the current flowing through the plasma chamber and of the electrode voltage are very close to sinusoidal form (figure 4.13). The phase angle of 88° with which the current leads the voltage (figure 4.13a) suggests the capacitive behavior of the chamber in the absence of the plasma. The decrease in the current to voltage phase angle when the plasma is ignited (figure 4.13b) is due to the resistance of the plasma bulk and it is used in the electrical diagnostic of the discharge.

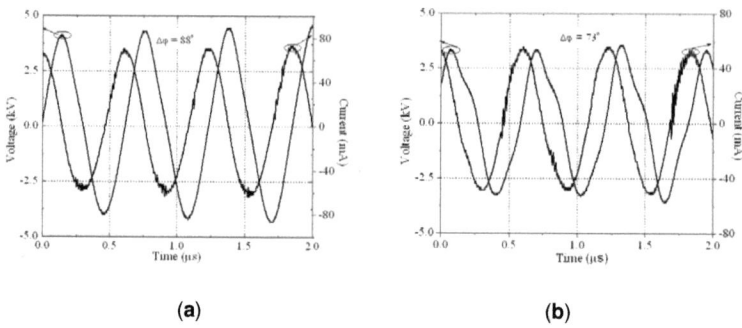

(a) (b)

Figure 4.13 – Electrodes voltage and current waveforms in the absence (**a**) and presence (**b**) of the homogeneous plasma (gas flow-rate, 1.1 l/min). [6]

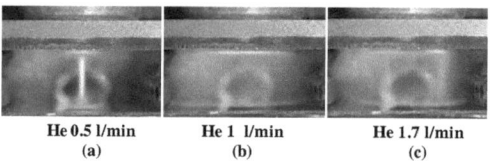

He 0.5 l/min   He 1 l/min   He 1.7 l/min
  (a)   (b)   (c)

Figure 4.14 – Plasma appearance as function of helium flow-rate. [6]

The discharge is initiated at a helium flow-rate of 0.5 l/min when the light emission from the plasma is visually observed. Initially the plasma has a thin column aspect and covers only small areas of the glass dielectrics. The plasma column is surrounded by a region of reduced brightness (figure 4.14a). With the increase of the helium flow-rate at 1 l/min, it spreads on the dielectrics taking a cylindrical shape with the same transversal area as the powered electrode. The discharge has a white-purplish glowing color and it appears uniform and homogeneous throughout the volume (plasma bulk) and as two dark spaces (plasma sheaths) near the dielectrics (figure 4.14b). A further increase in the gas flow-rate slightly raises the plasma emission intensity and causes a brighter emission at the bulk ends, near the dielectrics, than in the median bulk. At the same time, the thickness of the sheaths slightly increases. When the gas flow rate is raised above a critical level (1.7 l/min), the discharge turns into two or more thin brighter columns inside a region of reduced brightness. At a gas flow-rate higher than 2 l/min the plasma extinguishes (figure 4.14c).

The results of the electrical analysis (see section 3.2) are exhibited in Table 4.3. They are consistent with other results reported for radiofrequency AP helium DBD.

Table 4.3. Electrical characteristics of 1.74 GHz DBD.

R_{bulk} (kΩ)	I_{rms} (mA)	j_e (mA/cm^2)	n_e (cm^{-3})	P_{plasma} (W)	p_{plasma} (W/cm^3)
32.2	21.8	45.4	$1.1\ 10^{12}$	15.3	96.8

The gas kinetic temperature of DBD working in the homogeneous stage (power, 15.3 W and gas flow-rate, 1.1 l/min) measured with the thermocouple was of 119 °C. Knowing that at atmospheric pressure the rotational temperature of the molecules could be a good approximation of the gas kinetic temperature, the measured temperature was compared with the rotational temperature of the nitrogen ionic molecule. It was estimated to be of 124 °C. It is very close to the temperature measured with the thermocouple. This temperature value was confirmed by the softening of a PET (Polyethylene Terephthalate) foil introduced in the plasma. It is known that the heat distortion point of PET is 70 °C and the melting point is higher than 250 °C.

For comparison, a compact metallic disk-shape electrode placed between two glass plates was made. The electrode diameter and the glass plates have the same dimensions as in the case of the flat spiral electrode. Under the same working conditions as in figure 4.14b, the luminous area on the plastic surface is not uniform. The visual aspect is like three Newton's diffuse rings. The inner luminous area corresponds to the center of the disk-shape electrode and the outer luminous area has a diameter slightly higher than the electrode diameter. The area between them is less luminous. Our supposition is that the outer luminous area is determined by the more intense electric field generated at the electrode edge.

4.3

Plasma in contact with liquids

Because the plasma generated in liquids or on the liquid surface is currently being studied by us, information on their generation and possible applications will be summarized. By using the combination high voltage generator + modified fly-back transformer and a torch-like reactor (figures 2.1 and 2.4) plasma can be generated either on liquid or inside liquid.

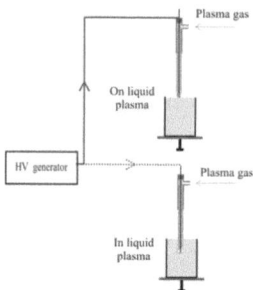

Figure 4.15 – Generation of APP on liquid or in liquid.

The estimated parameters of the plasma generated in Ar bubbled in water and at the water surface (figure 4.15) are presented in Table 4.4.

Table 4.4. Characteristics of Ar plasmas generated in and on water.

		In water	On water
T_{excAr}	[K]	6080	7510
T_{rotOH}	[K]	1990	2600
T_{vibrN_2}	[K]	-	2970
n_e	[cm^{-3}]	$3.79 \cdot 10^{15}$	$2.73 \cdot 10^{15}$

All parameters of the plasma generated in contact with liquids have values higher than those of the plasma micro-jet and of the dielectric barrier discharge at atmospheric pressure. This is important when the plasma is used for degradation of organic dyes or for nanostructures synthesis.

4.4

Plasma optical emission

Optical emission of plasma is largely used for its diagnosis by non-intrusive techniques in order to have information on plasma temperatures, electron number density or electrons energy. In is also a very useful indicator of chemically active species generated by the plasma which are very important in various plasma technologies, especially in the surfaces treatment or bacterial deactivation.

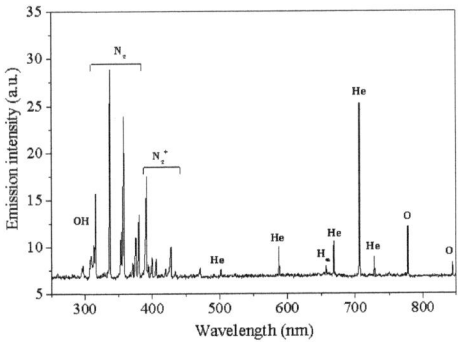

Figure 4.16 – Emission spectrum of atmospheric pressure helium DBD.

A typical emission spectrum of atmospheric pressure helium DBD in the range of 200-900 nm is shown in figure 4.16. The emission spectra of helium plasma jet and of helium DBD are very similar with the exception of relative intensities of atomic emission lines and of molecular emission bands (vibrational and rotational) and of the presence of relative intense emission bands of nitrogen oxide in the emission spectrum of atmospheric pressure

53

plasma jet (200 – 300 nm). They are due to the chemical conversion of N_2 and O_2. Beside the emission lines of helium, atomic emission lines of oxygen (777.41 and 844.67 nm) and hydrogen (656.27 nm, H_α) and molecular bands of OH, N_2 and N_2^+ are presented in the radiation spectrum.

With the exception of helium, the presence in the plasma of the other atomic and molecular species is inevitable because of the back diffusion of the ambient air. Recently it was demonstrated that small level of impurities (particularly nitrogen from ambient air) could have an important influence in the radiation of noble gas plasmas. The presence of free radicals and ions in the plasma could be important in the applications based on the generation of chemically active species. The OH radicals (emission band at 308.9 nm) represent the result of the dissociation of H_2O molecules from the humid back diffused air caused by the collisions with accelerated electrons or with long life species presented in the plasma, especially with helium metastables, He_m^*. Starting with the wavelength of 315 nm the UV spectrum is dominated by the emission of nitrogen molecules, the most representative being: the emissions of 2nd positive system of N_2 (315.93, 337.13, 357.69 and 380.49 nm) and of 1st negative system of N_2^+ (391.44 nm). It must be mentioned here that the N_2^+ emission is attributed to Penning ionization of N_2 with helium metastables. Otherwise, it is well known that nitrogen molecules are very effective at quenching the helium metastables. The presence of the basic atomic line of hydrogen, H_α, is due to the excitation of hydrogen atoms generated by the dissociation of H_2O molecules under the action of the energetic electrons in the plasma.

The addition of small quantities of argon (0.1–0.2 l min^{-1}) in helium could have the following effects: (**a**) the helium emission decreases drastically and is dominated by the emission of argon; (**b**) the electron number density, electron energy and characteristic temperatures increase; (**c**) plasma becomes smaller in volume and has a thinner visual aspect; and (**d**) the discharge could become unstable or even filamentary.

Figure 4.17 shows the emission spectrum of an Ar plasma generated in water. The plot is the result of the average of more scans in order to compensate the light fluctuation during an ignition-extinction discharge cycle. With the exception of the argon emission lines, the emission lines of hydrogen (H_α) and oxygen, and the emission bands of OH radical as a result of water molecules dissociation are present in the radiation spectrum.

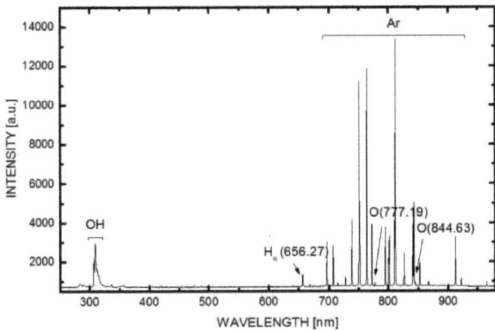

Figure 4.17 – Emission spectrum of plasma generated in Ar bubbled in water.

5 Potential applications [7,8]

This section presents in short possible applications of thermally non-aggressive plasmas generated with our laboratory-made devices. Because the generated plasmas have very low temperature they are tolerated by the thermo-sensitive material, being important for the chemically active species generated by processes which take place in the discharge. Our plasmas were tested for the activation of solid surfaces, bacterial deactivation, organic dye degradation and nanoparticle synthesis.

5.1
Bacterial deactivation and wettability improving with plasma jet

The effect of treatment with 48 kHz helium plasma-jet of tracing paper, laser-jet printing paper and PET foil was evaluated by measuring the contact angle of 1 μl distilled water droplet placed on the treated surface. The treatment time was varied in the range of 1 to 15 s. Figure 5.1a shows the enhancement of hydrophilicity of the treated materials. The effect becomes considerable after the first 4-5 s of treatment, showing saturation tendencies for longer treatment times.

Wettability changes for the printing paper, due to plasma treatment, were also evaluated by estimating the absorbed quantity of distilled water. After plasma treatment a droplet of 10 μl distilled water was placed on the paper square (1 cm^2) for a time of 10 s, after which the excess non-absorbed water was removed. The wettability of paper was estimate by means of the difference between the paper squares weight after and before plasma treatment. Figure 6.1b shows that the absorbed quantity of water increases with the treatment time by a factor of about 1.85. This finding is consistent with the conclusions drawn from contact angle measurements.

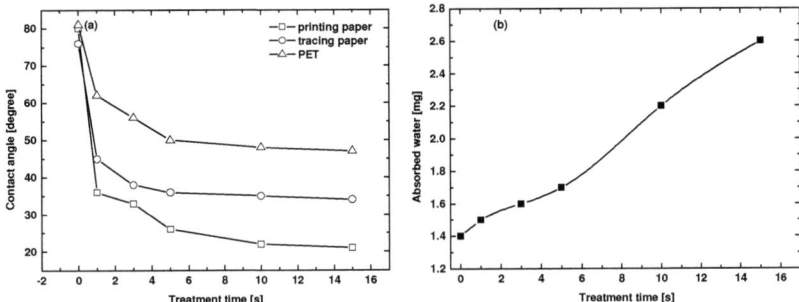

Figure 5.1 – Contact angle (**a**) and absorbed water (**b**) as function of treatment time. Treated materials: laser-jet printing paper, tracing paper and PET foil. [8]

The sterilizing capability of the helium plasma-jet was estimated by exposing microbial cultures, deposited on a glass plates, to the action of the plasma as function of treatment time. The methodology of preparing and handling the microbiological samples was the standard one. Figure 5.2 shows survival curves for microbial cultures of *B. subtilis* and *E. coli*. As it can be seen, the overall decimal time (treatment interval necessary to inactivate 90 % of the initial CFUs - colony forming units) is higher than 90 s.

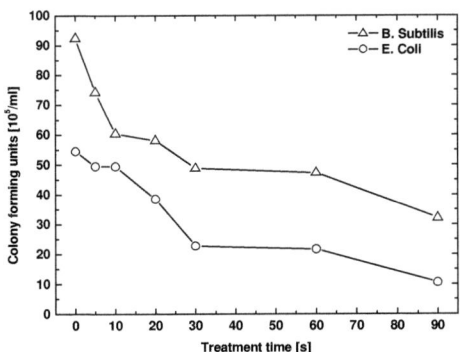

Figure 5.2 – Survival curves for *B. subtilis* and *E. coli*. [8]

For comparison, 718 kHz helium micro-jet was also tested as deactivation agent for *E. coli* cultures, the overall decimal time being around 25 s. This can be explained by the higher energy of electrons at higher working frequencies and as a consequence of the more chemically active species.

It was demonstrated that the deactivation capability of thermally non-aggressive plasma is better than other classical sterilization techniques such as dry heat or UV radiation techniques.

5.2
Surface treatment and sterilization with DBD

The presence of active species in 48 kHz DBD with spiral flat powered electrode was evidenced by testing the effect of the plasma on the paperboard hydrophilicity and on the transparency film wettability. Figure 5.3 shows the image of a piece of white paperboard (40x80 mm^2) on which droplets of 1 μl of commercial black ink were deposited in symmetric positions. The image was taken after 5 min from the treatment and deposition. The treatment duration was of 15 s. As it can be seen the diameter of the surface influenced by the plasma is at least 35mm, approximately equal to diameter of the plasma at the contact surface with the bottom wall of the plasma chamber.

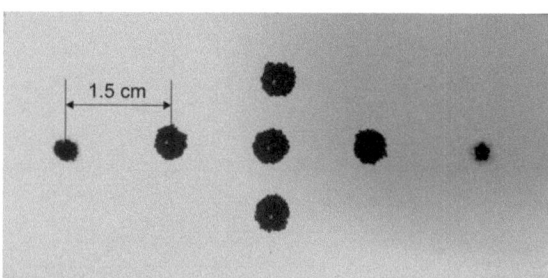

Figure 5.3 – The effect of treatment with 48 kHz DBD (spiral flat electrode) on the white paperboard hydrophilicity. [7]

The images of the 0.5 μl droplets of black ink deposited on the transparency films not exposed to the plasma, and exposed for 30 sec and 3 min respectively are presented in figure 5.4a-c.

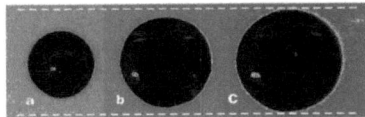

Figure 5.4 – The effect of treatment with 48 kHz DBD (spiral flat electrode) for 30 sec (**b**) and 3 min (**c**) on the transparency film wettability; (**a**) - untreated sample [7]

By using the spiral flat electrode powered with 48 kHz high voltage electric field larger volumes of helium plasma can be generated in closed spaces without a grounded electrode (metallic plate). In this way, the plasma generating system can also be used to treat the inner surfaces of flexible objects, such as small polymer bags (figure 5.5).

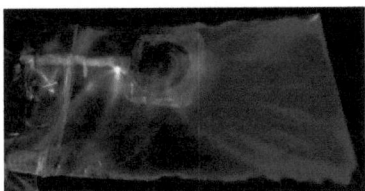

Figure 5.5 – Plasma can be generated in closed spaces without a grounded electrode. The polymer bag dimensions: 110 mm length and 70 mm width. Spiral flat electrode, 48 kHz, helium flow-rate: 3.4 l/min.

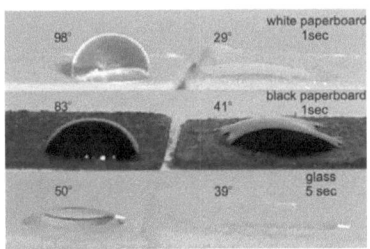

Figure 5.6 – Photographs and contact angles of water droplets deposited on different untreated and treated surfaces. 1.74 MHz helium DBD was used for treatment.

The possibility of using the 1.74 GHz helium DBD in the surface treatment was tested by observing the effect of the plasma on the paper and glass wettability. It was evaluated by observing and measuring the contact angle of 2 μl distilled water droplet placed on the treated surface. As it can be seen (figure 5.6), after only one second of exposure time, the effect of the plasma treatment (mainly on the paperboard wettability) is very visible.

5.3
Applications of in liquid plasma

Preliminary tests prove that the plasma generated in liquid as was shown in the section 3, could have at least three main applications: degradation of organic dyes and sterilization, nanoparticles synthesis and chemical analysis based on AES. Photographs presented in figures 5.7, TEM images from figure 5.8 and emission spectrum shown in figure 5.9 support the previous statements.

Figure 5.7 – Degradation of methylene blue by treatment with in liquid plasma.

The discoloration of methylene blue solution proves the existence of energetic electrons able to break the double bond of nitrogen separating the aromatic rings from its molecular structure and of hydroxyl radical with a strong oxidation character. Absorption measurements show that the energy efficiency is 8.67×10^{-10} mol J^{-1}. A secondary result of the treatment with plasma of the methylene blue solution, not mentioned so far in papers, is the progressive generation of solid particles (probably carbon structures) during the process. Their dimensions increase during the treatment time. It is noteworthy that recently this plasma was successfully tested for gold nanoparticles synthesis by using a new methodology.

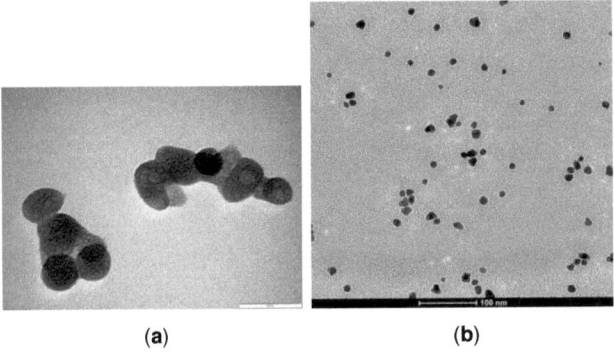

(a)   (b)

Figure 5.8 – TEM images of carbon nanostructures (**a**) and of gold nanoparticles (**b**) obtained by in liquid plasma treatment.

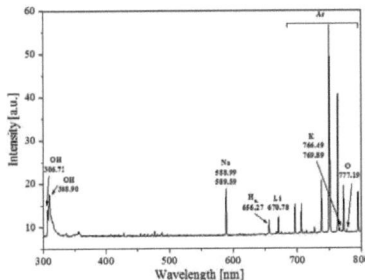

Figure 5.9 – Emission spectrum of plasma generated in an aqueous solution containing Na, Li and K.

At the surface contact between the plasma and the liquid vaporization, dissociation and atomization processes take place and many of generated atoms are excited on higher energetic states. So, the plasma in contact with liquids could be used for chemical analysis via AES. The emission spectrum of argon plasma generated in an aqueous solution containing Na, Li and K in concentration of 10 μg ml^{-1} confirms this. With the exception of argon emission lines, the emission lines of the three elements are well visible. There can also be observed the emission lines of hydrogen (H_α) and oxygen, and the emission bands of OH radical as a result of water molecules dissociation. These chemically active species generated by the plasma play an important role in organic dyes degradation, sterilization and nanostructures synthesis.

Finally, we have to mention that these applications of plasma generated in contact with liquids are under study.

References

1. Anghel S.D. "Non-thermal plasmas in resonant circuits" *Ed. Presa Universitara Clujeana, Cluj-Napoca* 2011.

2. Anghel S.D. and Simon A. "An alternative source for generating atmospheric pressure non-thermal plasmas" *Plasma Sourc. Sci. Technol.* 16, pp. B1-B4, published 3 August 2007. © IOP Publishing. Reproduced with permission. All rights reserved.

3. Anghel S. D. and Simon A. "Measurement of electrical characteristics of atmospheric pressure non-thermal He plasma" *Meas. Sci. Technol.* 18, pp. 2642-2648, published 11 July 2007. © IOP Publishing. Reproduced with permission. All rights reserved.

4. Anghel S.D., Frentiu T. and Simon A. "Atmospheric Pressure Plasmas in Resonant Circuits" *The Open Plasma Physics Journal* 2, 8-16, 2009.

5. Anghel S.D. "Generation and Electrical Diagnostic of an Atmospheric-Pressure Dielectric Barrier Discharge" *IEEE Trans. Plasma Sci.* 39, 871-876, 2011.

6. Reprinted from *Journal of Electrostatics,* vol.69, Anghel S.D., "Generation and investigation of a parallel-plate DBD driven at 1.6 MHz with flowing helium", pp. 261-264, © 2011 with permission from Elsevier.

7. Reprinted from *Journal of Electrostatics,* vol.71, Anghel S.D., "Atmospheric pressure plasma with a flat spiral electrode", pp. 155-158, © 2013 with permission from Elsevier.

8. Anghel S.D., Simon A., Papiu M.A., Dinu O.E. "A very low temperature atmospheric-pressure plasma jet in a single electrode configuration" *Roum. Journ. Phys. Suppl.* 56, 90-94, 2011.

9. Anghel S.D. and Simon A. "Plasma de înaltă frecvenţă", *Ed. Napoca Star*, 2002.

10. Bruggeman P. and Leys C. "Non-thermal plasmas in and in contact with liquids" *J. Phys. D: Appl. Phys.* 42, 053001, 2009.

11. Chen F.F. "*Introduction to plasma physics and controlled fusion. Volume 1, Plasma physics*" *Springer*, 2006.

12. Chirokov A., Gutsol A. and Fridman A. "Atmospheric pressure plasma of dielectric barrier discharges" *Pure Appl. Chem.* 77, 487-495, 2005.

13. Denicolai M. "Optimal performance for Tesla transformers" *Rev. Sci. Instrum.* 73, 3332-3336, 2002.

14. Fridman A., Chirokov A. and Gutsol A. "Non-thermal atmospheric pressure discharges" *J. Phys. D: Appl. Phys.* 38, R1-R24, 2005 .

15. Fridman A. and Kennedy L. A. "*Plasma Physics and Engineering*" *Ed. New York London Taylor & Francis*, 2004.

16. Fridman A. and Kennedy L. A. "Plasma Physics and Engineering" *Ed. New York London Taylor & Francis*, 2004.

17. Golubovskii Y.B., Maiorov V.A., Li P. and Lindmayer M. "Effect of the barrier material in a Townsend barrier discharge in nitrogen at atmospheric pressure" *J. Phys. D: Appl. Phys.* 39, 1574-1583, 2006.

18. Goossens O., Vangeneugden D., Paulussen S. and Dekempeneer E. "Physical and chemical properties of atmospheric pressure polymer films". Available: http://www.ut.ee/hakone8/papers/T7/Goossens.pdf.

19. Graham W. G. and Stalder K. R. "Plasma in liquids and some of their applications in nanoscience" *J. Phys. D: Appl. Phys.* 44, 174037, 2011.

20. Janca J., Zajickova L., Klima M. and Slavicek P. *Plasma Chem. Plasma Proc.* „Diagnostics and Application of the High Frequency Plasma Pencil" 21, 565-579, 2001.

21. Johnson G.L., "Tesla Coil Impedance". Available: http://www.eece.ksu.edu/~ Gjohnson /TeslaCoilImpedance.pdf

22. Kanazawa S., Gogoma M. Moriwaki T. and Okazaki S. "Stable glow plasma at atmospheric pressure" *J. Appl. Phys. D: Appl. Phys.* 21, 838-840, 1988.

23. Kieft I.E., v d Laan E.P. and Stoffels E. "Electrical and optical characterization of the plasma needle" *New J. Phys.* 149, 1-14, 2004.

24. Kogelschatz U. "Filamentary, Patterned, and Diffuse Barrier Discharges" *IEEE Trans. Plasma Sci.* 30, 1400-1408, 2002.

25. Laimer J. and Stori H. "Glow Discharges Observed in Capacitive Radio-Frequency Atmospheric-Pressure Plasma Jets" *Plasma Process. Polym.* 3, 573-586, 2006.

26. Laimer J., Haslinger S. and Störi H. "Characterization of an Atmospheric Pressure Radio-Frequency Capacitive Plasma Jet" *Plasma Process. Polym.* 4, S487-S492, 2007.

27. Laroussi M., Alexeff I., Richardson J.P. and Dier F F. "The Resistive Barrier Discharge" *IEEE Trans. Plasma Sci.* 30, 158-159, 2002.

28. Laroussi M., Tendero C., Lu X., Alla S. and Hynes W.L. „Inactivation of Bacteria by the Plasma Pencil" *Plasma Process. Polym.* 3, 470-473, 2006.

29. Law V.J., Milosavljević V., O'Connor N., Lalor J.F. and Daniels S. "Handheld Flyback driven coaxial dielectric barrier discharge: "Development and characterization" *Rev. Sci. Instrum.,* 79, 094707, 2008.

30. Law J.V. and Anghel S.D. *"Compact atmospheric pressure plasma self-resonant drive circuits" J. Phys. D: Appl. Phys.* 45, 075202, published 1 February 2012. © IOP Publishing. Reproduced with permission. All rights reserved.

31. Léveillé V. and Coulombe S. „Design and preliminary characterization of a miniature pulsed RF APGD torch with downstream injection of the source of reactive species" *Plasma Sources Sci. Technol.* 14, 467-476, 2005.

32. Léveillé V. and Coulombe S. „Electrical probe calibration and power calculation for a miniature 13.56 MHz plasma source" *Meas. Sci. Technol.* 17, 3027-3032, 2006.

33. Lieberman M.A. "Dynamics of a collisional, capacitive RF sheath" *IEEE Trans. Plasma Sci.* 17, 338-341, 1989.

References

34. Lieberman M.A. "Principle *of plasma discharges and material processing*" Ed. *John Wiley & Sons*, Inc., 1994.

35. Luque H. and Crosley D.R. "Lifbase" *Report No. MP 99-009* Stanford Research Institute (SRI), 1999. Available: http://www.sri.com/cem/lifbase.

36. Mariotti D. and Sankaran R. M. "Perspectives on atmospheric-pressure plasmas for nanofabrication" *J. Phys. D: Appl. Phys.* 44, 174023, 2011.

37. Moon S.Y., Kim D.B., Gweon B. and Choe W. "Driving frequency effects on the characteristics of atmospheric pressure capacitive helium discharge" *Appl. Phys. Lett.* 93, 221506 1-3, 2008.

38. Nastuta A.V., Rusu G.B., Topala I., Chiper A.S. and Popa G. "Surface modifications of polymer induced by atmospheric DBD plasma in different configurations" *J. Optoelectron. Adv. Mater.* 10, 2038-2042, 2008.

39. Navratil Z., Trunec D., Smid R. and Lazar L. "A software for optical emission spectroscopy-problem formulation and application to plasma diagnostics" *Czech. J. Phys.* 56B, 944-951, 2006.

40. Niemi K., Wang Sh, Schultz-von der Gathen V. and Döbele H. F. *Poster Conference Frontiers on Low Temperature Plasma Diagnostics* Lecce, Italy, 2003. Available online: http://fltpd-5.ba.cnr.it/Paper/Libro/239.pdf

41. Ostrikov K., Cvelbar U. and Murphy A. B. "Plasma nanoscience: setting directions, tackling grand challenges" *J. Phys. D: Appl. Phys.* 44, 174001, 2011.

42. Park J., Henins I., Herrmann H.W. and Selwyn G.S. "Discharge phenomena of an atmospheric pressure radio-frequency capacitive plasma source" *J. Appl. Phys.* 89, 20-28, 2001.

43. Park J., Henins I., Herrmann H.W., Selwyn G.S., Jeong J.Y., Hicks R.F., Shim D. and Chang C.S. "An atmospheric pressure plasma source" *Appl. Phys. Lett.*, 76, 288-290, 2000.

44. Raizer Y.P. *"Gas Discharge Physics"* Ed. *Berlin Heidelberg Springer-Verlag*, 1991.

45. Rasband W.S. "ImageJ. Image Processing and Analysis in Java", U.S. National Institutes of Health, Bethesda Maryland USA. Available: http://rsb.info.nih.gov/ij/.

46. Roth J.R., Sherman D.M., Ben Gadri R., Karakaya F., Chen Z., Montie T.C., Kelly-Winterberg K. and Tsai P. P-Y. "A remote exposure reactor (RER) for plasma processing and sterilization by plasma active species at one atmosphere" *IEEE Trans. Plasma Sci.* 28, 56-63, 2001.

47. Schutze A., Jeong J.Y., Babayan S.E., Park J., Selwin G.S. and Hicks R.F. "The atmospheric-pressure plasma jet: a review and comparison to other plasma sources" *IEEE Trans. Plasma Sci.* 26, 1685-1694, 1998.

48. Shi J. J., Liu D. W. and Kong M. G. "Effects of Dielectric Barriers in Radio-Frequency Atmospheric Glow Discharges" *IEEE Trans. Plasma Sci.*, 35, 137-142, 2007.

49. Shi J.J., Liu D.W. and Kong M.G. "Plasma stability control using dielectric barriers in radio-frequency atmospheric pressure glow discharges" *Appl. Phys. Lett.* 89, 081502 1-3, 2006.

50. Reprinted from *Journal of Electrostatics,* vol.70, Simon A., Dinu O.E., Papiu M.A., Tudoran C., Papp J. and Anghel S.D., "A study of 1.74 MHz atmospheric pressure dielectric barrier discharge for non-conventional treatments", pp. 235–240, © 2012 with permission from Elsevier.

51. Stoffels E., Flikweert A.J., Stoffels W.W. and Kroesen G.M.W. "Plasma needle: a non-destructive atmospheric plasma source for fine surface treatment of (bio)materials" *Plasma Sources Sci. Technol.* 11, 383-388, 2002.

52. Tudoran C.D. "Simplified portable 4 MHz RF plasma demonstration unit" *Journal of Physics: Conference Series,* 182, 012034 1-4, 2009.

53. Uglum J.R. "Analysis of the Resonant Transformer Accelerator in the Gated and Ungated Beam Mode" *IEEE Trans. Nucl. Sci.* 3, 1026-1029, 1975.

54. Walsh J.L. and Kong M.G. "10 ns pulsed atmospheric air plasma for uniform treatment of polymeric surfaces" *Appl. Phys. Lett.* 91, 251504 1-3, 2007.

55. Walsh J.L., Shi J.J. and Kong M.G. "Contrasting characteristics of pulsed and sinusoidal cold atmospheric plasma jets" *Appl. Phys. Lett.* 88, 171501 1-3, 2006.

56. Yang S. and Yin H. "Two Atmospheric-pressure Plasma Sources for Polymer Surface Modification" *Plasma Chem. Plasma Process.* 27, 23-33, 2007.

Appendix [1]

Laplace transform

The *Laplace transform* provides a method for representing and analyzing linear systems using algebraic methods. AC circuit analysis may be conducted in the time domain with differential equations or in the so-called *complex frequency domain*. By working in the frequency domain, the operations of integration and differentiation in the time domain become simple algebraic expressions. The *Laplace transform* converts a problem between these two domains, time and frequency. It is used for circuit analysis, facilitating the calculation of a network time response and represents a huge improvement over working directly with differential equations. The Laplace transform of the output signal, $Y(s) = \mathcal{L}[y(t)]$ (figure A1) is simply the product of the Laplace transform of the input signal, $X(s) = \mathcal{L}[x(t)]$ and the Laplace transform of the network transfer function, $H(s) = \mathcal{L}[h(t)]$. The time response of the network will be found as the Inverse Laplace transform of this product, $y(t) = \mathcal{L}^{-1}[Y(s)]$.

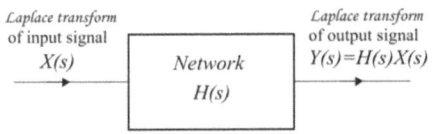

Laplace transform of input signal
$X(s)$

Network
$H(s)$

Laplace transform of output signal
$Y(s) = H(s)X(s)$

$H(s)$ - *Laplace transform* of the network transfer function

Figure A1

The Laplace transform characterizes the exponential response of a time-invariant linear function. The **bilateral** *Laplace transform* of a continuous-time function $x(t)$ is:

$$\mathcal{L}[x(t)] = X(s) = \int_{-\infty}^{\infty} x(t)e^{-st}dt$$

where $s = \sigma + j\omega$ is called the *complex frequency*. The special case $\sigma = 0$ corresponds to $s = j\omega$, that is the steady-state sinusoidal case.

When the function $x(t) = 0$ is defined only for $t \geq 0$ the relationship:

$$\mathcal{L}[x(t)] = X(s) = \int_0^\infty x(t)e^{-st}dt$$

is known as **unilateral** Laplace transform.

It is said about $x(t)$ and $X(s)$ that they form a Laplace transform pair. $X(s)$ is the Laplace transform of $x(t)$, and $x(t)$ is the inverse Laplace transform of $X(s)$.

$X(s) = \mathcal{L}[x(t)]$ and $x(t) = \mathcal{L}^{-1}[X(s)]$

where:

$$\mathcal{L}^{-1}[X(s)] = \int_0^{j\infty} X(s)e^{st}ds$$

Very often, to solve problems involving Laplace transform, conversion from t domain to s domain and vice versa is done via the correspondence table (Table A1) and based on the general properties of Laplace transform (Table A2). Thus, the integration and differentiation operations are avoided.

Table A1 - Main used Laplace transform pairs

Signal $x(t)$ t domain	Laplace Transform $X(s)$ s domain
$\delta(t)$ unit impulse	1
$u(t)$ unit step	$\dfrac{1}{s}$
$u(t)\,t$	$\dfrac{1}{s^2}$
$u(t)\,e^{-at}$	$\dfrac{1}{s+a}$
$u(t)\sin \omega t$	$\dfrac{\omega}{s^2 + \omega^2}$
$u(t)\cos \omega t$	$\dfrac{s}{s^2 + \omega^2}$
$u(t)\,e^{-at}\sin \omega t$	$\dfrac{\omega}{(s+a)^2 + \omega^2}$
$u(t)\,e^{-at}\cos \omega t$	$\dfrac{s+a}{(s+a)^2 + \omega^2}$
$u(t)\sin(\omega t + \varphi)$	$\dfrac{s\sin \varphi + \omega \cos \varphi}{s^2 + \omega^2}$

$u(t)\cos(\omega t + \varphi)$	$\dfrac{s\cos\varphi - \omega\sin\varphi}{s^2 + \omega^2}$
$u(t)\dfrac{1}{a}\left(1 - e^{-at}\right)$	$\dfrac{1}{s(s + a)}$
$u(t)\dfrac{a - (a - b)e^{-bt}}{b}$	$\dfrac{s + a}{s(s + b)}$
$u(t)\dfrac{1}{b - a}\left(e^{-at} - e^{-bt}\right)$	$\dfrac{1}{(s + a)(s + b)}$
$u(t)\dfrac{1}{b - a}\left(be^{-bt} - ae^{-at}\right)$	$\dfrac{s}{(s + a)(s + b)}$

Table A2 - Main used Laplace transform properties

Property	Signal $x(t)$ t domain	Laplace Transform $X(s)$ s domain
Linearity	$\alpha_1 x_1(t) + \alpha_2 x_2(t)$	$\alpha_1 X_1(s) + \alpha_2 X_2(s)$
Time differentiation	$\dfrac{dx}{dt}$	$sX(s)$
Time integration	$\displaystyle\int_{-\infty}^{t} x(\tau)d\tau$	$\dfrac{1}{s}X(s)$
Time shift	$x(t - t_o)$	$e^{-st_o}X(s)$
Time convolution	$\displaystyle\int_{-\infty}^{+\infty} h(\tau)x(t - \tau)dt$	$H(s)X(s)$

To facilitate the application of Laplace transform in analyzing practical circuits working in pulse regime, consider a sinusoidal signal:

$$u(t) = U\,e^{j\omega t}$$

whose derivative is.

$$\frac{du(t)}{dt} = j\omega U\,e^{j\omega t} = j\omega u(t)$$

The transition from t (time) domain to s (frequency) domain is done by the substitution:

$$j\omega = s$$

$$\frac{d}{dt}u(t) = s\,u(t) \quad \Rightarrow \quad \frac{d}{dt} \rightarrow s$$

As it can be seen, the Laplace transform substitutes the differential operator d/dt by the Laplace transform operator, s. This substitution transforms differential equations in algebraic equations where the s term may be manipulated like any other variable. It is very useful in analyzing electrical networks without solving differential equations. Table A3 shows the correspondence between the impedances of the basic circuit elements in t domain and s domain respectively, and Table 4 shows how they are used in electric diagrams.

Table A3 - Impedances in t and s domain of main used circuit elements

	Differential equation t domain	Impedance t domain	Algebraic equation s domain	Impedance s domain
Capacitor	$i_C(t) = C\dfrac{d}{dt}u_c(t)$	$\dfrac{1}{j\omega C}$	$i_c(s) = Cs\,u_c(s) = sC\,u_c(s)$	$\dfrac{1}{sC}$
Inductor	$u_L(t) = L\dfrac{d}{dt}i_L(t)$	$j\omega L$	$u_L(s) = Ls\,i_L(s) = sL\,i_L(s)$	sL
Resistor	$u_R(t) = i_R(t)R$	R	$u_R(s) = i_R(s)R$	R

Table A4 - Correspondence between main circuit elements in t and s domain in electric diagrams.

	Circuit element t domain	Circuit element s domain
Constant voltage source	E	$\dfrac{E}{s}$
Constant current source	I	$\dfrac{I}{s}$
Resistor	$u_R(t)$ R $i_R(t)$	$u_R(s)$ R $i_R(s)$
Capacitor (charged)	$u_c(t)$ $\dfrac{1}{j\omega C}$ $i_c(t)$ $u_c(0)=U_o$	$u_c(s)$ $i_c(s)$ $\dfrac{U_o}{s}$ $\dfrac{1}{sC}$

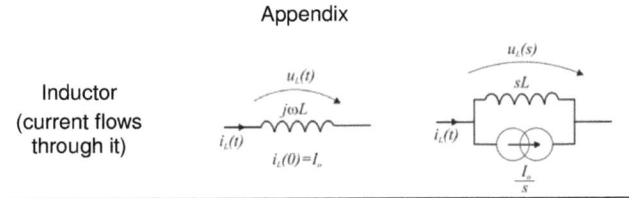

Inductor
(current flows
through it)

The process of applying Laplace transform to analyze a circuit involves the following steps:

(a) Converting the circuit in its Laplace equivalent by replacing each element of the circuit with its Laplace equivalent.

(b) Writing the algebraic equations by applying electrical networks theorems.

(c) Solving the equations system.

(d) Converting the solutions from *s* domain to *t* domain based on correspondence table.

Example

In the circuit shown in figure A2a, K switches from 1 to 2 at t = 0. Find the time dependence of the current passing through the resistor R for $t \geq 0$.

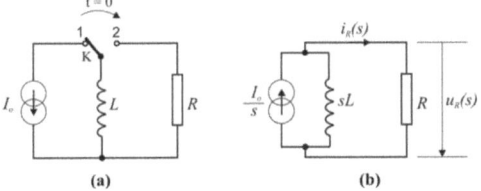

(a)   (b)

Figure A2

For t < 0, the current I_o flows through the inductor. For $t \geq 0$ the circuit in *s* domain is shown in Fig. A2b. The voltage (in *s* domain) across the resistor *R* is:

71

$$u_R(s) = \frac{I_o}{s} \frac{R \, sL}{R + sL} = I_o R \frac{1}{s + \dfrac{R}{L}}$$

and the current flowing through it is:

$$i_R(s) = I_o \frac{1}{s + \dfrac{R}{L}}$$

The inverse Laplace transform of

$$X(s) = \frac{1}{s + \dfrac{R}{L}}$$

is

$$x(t) = e^{-\frac{R}{L}t} \quad (\textit{Table A1.2}).$$

Finally, the time dependence of the current flowing through the resistor R has the expression:

$$\boxed{i_R(t) = I_o \, e^{-\frac{R}{L}t}}$$

i want morebooks!

Buy your books fast and straightforward online - at one of world's fastest growing online book stores! Environmentally sound due to Print-on-Demand technologies.

Buy your books online at
www.get-morebooks.com

Kaufen Sie Ihre Bücher schnell und unkompliziert online – auf einer der am schnellsten wachsenden Buchhandelsplattformen weltweit! Dank Print-On-Demand umwelt- und ressourcenschonend produziert.

Bücher schneller online kaufen
www.morebooks.de

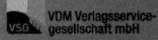 VDM Verlagsservice-
gesellschaft mbH

VDM Verlagsservicegesellschaft mbH
Heinrich-Böcking-Str. 6-8   Telefon: +49 681 3720 174   info@vdm-vsg.de
D - 66121 Saarbrücken   Telefax: +49 681 3720 1749   www.vdm-vsg.de

Printed by Books on Demand GmbH, Norderstedt / Germany